Hamlet's BlackBerry

:)

Hamlet's BlackBerry

A PRACTICAL PHILOSOPHY FOR BUILDING A GOOD LIFE IN THE DIGITAL AGE

WILLIAM POWERS

HARPER

An Imprint of HarperCollins*Publishers*

www.harpercollins.com

HarperCollins books may be purchased for educational, business, or sales promotional use. For information, please write: Special Markets Department, HarperCollins Publishers, 10 East 53rd Street, New York, NY 10022.

FIRST EDITION

Designed by Renato Stanisic

Library of Congress Cataloging-in-Publication Data has been applied for.

ISBN: 978-0-06-168716-7

10 11 12 13 14   ID/RRD   10 9 8 7 6 5 4 3 2 1

For Ann Shallcross,
who connected

This time, like all times, is a very good one, if we but know what to do with it.

—Ralph Waldo Emerson

Contents

Prologue

The Room

T*ap, tap, tap, tap, tap, tap. . .*

Imagine you're in a gigantic room, a room so spacious it can comfortably hold more than a billion people. In fact, that's how many people are there with you right now.

Despite its size, the room is ingeniously designed so everyone is in close proximity to everyone else. Thus any person in the room can easily walk over to any other person and tap him or her on the shoulder.

As you move around the room each day, this is exactly what happens. Wherever you go, people come up to you and tap you on the shoulder. Some tap gently, some firmly, but they all want the same thing: a little of your time and attention.

Some ask you questions and wait for answers. Others request favors. There are people who are eager to sell you things and other people who want to buy things of yours.

Some share personal news and photos from their recent travels. Others want only to talk business. Occasionally someone taps to say they miss you—which is a little odd, since they're right there in the room with you—and they give you big hugs and kisses. Certain friends tap often to keep you abreast of everything they're thinking and doing, no matter how trivial. "I'm now eating a cheeseburger," one will say and hold it up for you to see.

Encounters often overlap. As you're dealing with one person, another comes along and taps, and you have to choose between them.

You do pretty well managing all these overtures, while also making your own. It's kind of thrilling to be in the room. There's always something going on, and you're learning a lot. And some of these people—maybe twenty or thirty out of the billion—really matter to you. You make a point of tapping them as often as possible, and when they tap you back, it really feels nice.

Tap, tap, tap, tap, tap, tap.

So it goes all day and night. The enormous room is a nonstop festival of human interaction.

Like everyone else in the room, you have a personal zone where you eat, sleep, and hang out. Your zone is nicely furnished and quite comfortable. But it has no walls, and people seeking you out can come in at any time. If you happen to be asleep, they leave messages, sometimes labeling them URGENT. You find them when you wake up each morning, dozens of messages waiting for answers.

After a few years, you grow a bit tired of life in the room. It's getting exhausting, all this tapping. You crave some time away from all those other people, their needs and demands, and the strange pull that life in the room has on you.

So you decide to take a little vacation. You'll leave the room for a few days, go someplace where nobody can find you. You know exactly the kind of place it will be, too: fresh air, a big empty sky, no sound except the birds and the wind moving through the trees. Best of all, there will be no other people. You'll just sit by yourself and let your mind float free.

The more you imagine it, the more you can't wait to be there. Why didn't you think of this before?

You pack a small bag and head for the outer reaches of the room. After a short time, you come to one of the walls. Your eyes sweep its surface, searching for a door. There doesn't seem to be one, but the wall continues in both directions. For no particular reason, just a hunch, you turn left.

You walk on, following the perimeter of the room while watching carefully for an exit. As on any other day, people come over frequently and tap you. There's a new tap every few minutes.

After responding to each one's comment or query, you ask if they can direct you to the nearest door out of the room. You do this over and over, but nobody offers any helpful information. Most say they don't know of any doors and apologize that they can't help.

A few seem mildly put off by your question. They stare directly into your eyes for the briefest moment, as if you're a puzzle they'd like to solve.

Only one person, a young woman wearing a straw hat, seems truly happy you asked the question.

"A door?" she says. "I can't believe you asked me that. I've been wondering the same thing for years. If you find one, will you tap and let me know? I'd give anything to go outside for an hour."

You begin to ask what makes her say that, but before you can finish, she's interrupted by another woman tapping her from behind.

"Good luck!" she says, with a sweet smile and a wave. "Don't forget me!"

You continue walking. Hours go by, and still no door. This is strange. Before you moved to the room, taking off used to be so simple. When you were a kid, your parents would pile the whole family in the car and drive out to the lake. You'd spend two weeks together there in an old cabin and not hear from anyone.

After college, when you were living in the city, you used to get away almost every weekend, grab a friend and head down to the beach or up to the mountains. It wasn't complicated. Anyone could do it.

Finally, just when you're about to give up, you come to a large hole in the wall. There are people milling around nearby, but they're all turned away from the hole, as if they don't know it's there or know and don't care.

It's not exactly a door. It's an arch-shaped opening about ten feet high by four feet wide, with a ledge running along the bottom at about thigh level. The ledge is deep and flat, perfect for someone to sit on and contemplate the view. There's nobody sitting on it now.

You peer outward. The view isn't what you imagined. You expected mountains and broad valleys with roads winding lazily through them—the vistas of happy vacations. But all you can see is a black backdrop decorated with tiny white twinkle lights, like the lights people in the room string on their Christmas trees for the holidays.

After a few minutes your eyes adjust, and you realize those aren't twinkle lights at all. They're stars! You're looking out into space, the cosmos. It seems the room has broken free of the earth, the way a huge chunk of ice occasionally breaks off a glacier and floats away on its own. You remember reading about this once. They say the glacier has "calved." What do they say when it happens to a big room filled with people?

Your options are clear. You can turn around and go back to your personal zone, or you can step through the arch and see what happens.

The latter course is risky. Will you be able to breathe out there? Will you float pleasantly away from the room, or will it feel scary, more like falling?

Once outside, you'll want to find your way back to Earth, and to do that you'll probably need help. Will you meet others who have left the room before you and know the way?

It occurs to you that there's a chance nobody has ever tried this before. If they had, wouldn't you have heard something about it? News travels very quickly in the room.

As you're pondering all this, you feel a tap on your shoulder. Normally you'd turn right around and answer it. But this time you hesitate. Part of you is curious to know who's there and what he wants. Is it someone you know? A total stranger? But you're so transfixed by the view, you can't bear to turn away even for a moment. It's the first time since moving to the room that you've totally ignored a tap. It feels wild and, somehow, right.

You climb up so you're standing on the ledge, one hand holding the side of the arch for balance. You lean out a little to see what's below. More stars, endless stars.

You sense movement beside you.

"I hope you're not angry that I followed you," says a voice you recognize. The woman in the straw hat is clambering up.

"Here," you say, offering her your hand.

"Thank you," she says, now on her feet. "I couldn't resist. There's nothing I want more than this." She throws her arms out to the universe, like a singer belting out a song.

"Ready?" you ask, and she nods.

You close your eyes, bend your knees slightly, and leap!

Hamlet's BlackBerry

:)

Introduction

This book is about a yearning and a need. It's about finding a quiet, spacious place where the mind can wander free. We all know what that place feels like, and we used to know how to get there. But lately we're having trouble finding it.

Like the people in the story you just read, we live in a world where everyone is connected to everyone else all the time. We're not literally in a room that's floated away from the earth, but we're definitely in a new place, and it's technology that has brought us here. Our room is the digital space, and we tap each other through our connected screens.* Today we're always just a few taps away from millions of other people, from endless information and stimulation. Family and friends, work and play, news and ideas—sometimes it seems everything we care about has moved to the digital room. So we spend our days there, living in this new ultra-connected way.

We've been at it for about a decade now, and it's been thrilling and rewarding in many ways. When the whole world

---

I use the word "screens" here and throughout the book as shorthand for the connective digital devices that have been widely adopted in the last two decades, including desktop and notebook computers, mobile phones, e-readers, and tablets.

is within easy reach, there's no end of things to see and do. Sometimes it feels like a kind of a paradise.

However, there's a big asterisk to life in this amazing place. We've been doing our best to ignore it, but it won't go away. It comes down to this: We're all busier. Much, much busier. It's a lot of work managing all this connectedness. The e-mails, texts, and voicemails; the pokes, prods, and tweets; the alerts and comments; the links, tags, and posts; the photos and videos; the blogs and vlogs; the searches, downloads, uploads, files, and folders; feeds and filters; walls and widgets; tags and clouds; the usernames, passcodes, and access keys; pop-ups and banners; ringtones and vibrations. That's just a small sample of what we navigate each day in the room. By the time you read this there will be completely new modes of connecting that are all the rage. Our tools are fertile, constantly multiplying.

As they do, so does our busyness. Little by little, our workdays grow more crowded. When you carry a mobile device, all things digital (and all people) are along for the ride. Home life is busier too. Much of what used to be called free time has been colonized by our myriad connective obligations, and so is no longer free.

It's easy to blame all this on the tools. Too easy. These tools are fantastically useful and enrich our lives in countless ways. Like all new technologies, they have flaws, but at bottom they can't make us busy until we make them busy first. We're the prime movers here. We're always connected because we're always connecting.

Beyond the sheer mental workload, our thoughts have acquired a new orientation. Of the two mental worlds everyone inhabits, the inner and the outer, the latter increasingly rules. The more connected we are, the more we depend on the world outside ourselves to tell us how to think and live. There's always been a conflict between the exterior, social self and the interior, private one. The struggle to reconcile them is central to the human experience, one of the great themes of philosophy, literature, and art. In our own lifetime, the balance has tilted decisively in one direction. We hear the voices of others, and are directed by those voices, rather than by our own. We don't turn inward as often or as easily as we used to.

In one sense, the digital sphere is all about differentiating oneself from others. Anyone with a computer can have a blog now, and the possibilities for self-expression are endless. However, this expression takes place entirely *within* the digital crowd, which frames and defines it. This makes us more reactive, our thinking contingent on others. To be hooked up to the crowd all day is a very particular way to go through life.

For a long time, there was an inclination to shrug all of this off as a mere transitional issue, a passing symptom of technological change. These are early days, we tell ourselves. Eventually, life will calm down and the inner self will revive. There's a basic wisdom in this hopeful view. It's never a good idea to buy into the dark fears of the techno-Cassandras, who generally turn out to be wrong. Human beings are skillful at figuring out the best uses for new tools. However, it can take a while. The future is full of promise, but we have to focus on the present, how we're living, thinking, and feeling right now.

Like the two wayfarers in my story, a lot of us are feeling tapped out, hungry for some time away from the crowd. Life in the digital room would be saner and more fulfilling if we knew how to leave it now and then.

But *can* we leave? It's nice to imagine that there's a door somewhere and all you'd have to do is step through it and you'll be in a different place. A less connected place where time isn't so fugitive and the mind can slow down and be itself again. If someone told you that that place existed and he knew the way there, would you follow him?

What I'm proposing here is a new digital philosophy, a way of thinking that takes into account the human need to connect outward, to answer the call of the crowd, *as well as* the opposite need for time and space apart. The key is to strike a balance between the two impulses.

THE BOOK BEGINS with a brief look at the essential conundrum: Our screens perform countless valuable tasks for individuals and for businesses and other organizations. They deliver the world to us, bringing all kinds of convenience and pleasure. But as we connect more and more, they're changing the nature of everyday life, making it more frantic and rushed. And we're losing something of great value, a way of thinking and moving through time that can be summed up in a single word: depth. Depth of thought and feeling, depth in our relationships, our work and everything we do. Since depth is what makes life fulfilling and meaningful, it's astounding that we're allowing this to happen.

We've effectively been living by a philosophy, albeit an unconscious one. It holds that (1) connecting via screens is good, and (2) the more you connect, the better. I call it Digital Maximalism, because the goal is maximum screen time. Few of us have *decided* this is a wise approach to life, but let's face it, this is how we've been living.

There's an emerging recognition that this approach is causing us all kinds of problems. We sense it in our everyday lives—the constant need to check the screen, the inability to slow down our thoughts and focus. It's rampant at home, in school, and in the workplace. Various solutions have been proposed, ranging from behavioral regimens to software gadgets designed to help manage the flow of information. They haven't worked; the maximalist approach still rules.

What to do? Until recently, nobody has lived in a world of digital screens, so it would seem we are in uncharted territory. In fact, we're not. Human beings have been connecting across space and time, and using technology to do it, for thousands of years. And whenever new devices have emerged, they've presented the kinds of challenges we face today—busyness, information overload, that sense of life being out of control.

These challenges were as real two millennia ago as they are today, and throughout history, people have been grappling with them and looking for creative ways to manage life in the crowd. We can learn a great deal from their experience and the practical ideas that emerged from it. Though this book opened with a futuristic allegory, its premise is that the best place to find a new philosophy for a digital world—the door to a saner, happier life—is in the past.

In part II, I look at seven key moments from history, eras much like our own in their great technological ferment and also great confusion. In each period, I focus on one thinker who was unusually thoughtful about the tools of the time, tools that in many cases are still in use today. Their names are well known—Plato, Seneca, Gutenberg, Shakespeare, Franklin, Thoreau, and McLuhan—but their insights on this subject are less familiar.

Plato, for instance, shows that even in ancient Greece, people worried about what the latest technology was doing to their minds, and found ways to escape the crowd. Hamlet is one of literature's best-known characters, but you may not know that Shakespeare gave the Prince of Denmark a hot gadget, a handheld device that was as fashionable in Renaissance England as iPhones and BlackBerrys are today. The Seven Philosophers of Screens, as I call them, provide a tour of the technological past, focused on the human questions confronting us today. What do you do when your life has become too outward and crowd-driven? How to quiet the busy mind? For me, just knowing that these issues have come up so often before, under such different circumstances, is comforting and inspiring.

In the final part of the book, I offer guidelines for applying the lessons of the past, using real-world examples from today, along with a case study from my own life. The essential idea is simple: to lead happy, productive lives in a connected world, we need to master the art of disconnecting. Even in a world as thoroughly connected as ours, it's still possible to put some space between yourself and the crowd.

Humans love to journey outward. The connective impulse is central to who we are. But it's the return trip, back to the self and the life around us, that gives our screen time value and meaning. Why shouldn't we aim for a world that serves both needs?

The room is feeling kind of crowded, don't you think? Let's get away from it all.

PART I

WHAT LARKS?

The Conundrum of the Connected Life

Chapter One

BUSY, VERY BUSY

In a Digital World, Where's the Depth?

When I look around at how so many of us live today, staring into screens all the time, I think of my friend Marie. When I first met her in the mid-1990s, Marie was a recent immigrant to the United States and still learning the fine points of English. Back then, whenever I saw her and asked how she was doing, she would flash a big happy smile and say, "Busy, very busy!"

This was strange, partly because she said it so consistently and partly because her expression and upbeat tone didn't match her words. She seemed pleased, indeed ecstatic, to be reporting that she was so busy.

After a while, I figured out what was going on. Marie was copying what she'd heard Americans saying to one another over and over. Everyone talked so much about how busy they were, she thought it was a pleasantry, something that a person with good manners automatically said when a friend asked how they were doing. Instead of "Fine, thank you," you were supposed to say you were busy.

She was wrong, of course, as she eventually realized. But in another way she was absolutely right. "Busy, very busy" is exactly what we are most of the time. It's staggering how many balls we keep in the air each day and how few we drop. We're so busy, sometimes it seems as though busyness itself is the point.

What *is* the point, anyway? What's the goal at the bottom of all this juggling and rushing around? It's one of those questions you avoid thinking about because it's so hard to answer. When you start wondering about your own busyness, pretty soon you're pondering much deeper questions such as, Is this the kind of life I really want? From there it's just a short hop to the big-league existential stumpers, Why are we here? and Who am I?

Few of us are eager to take on such questions, and even if we were, who has time? We're all too busy! Besides, at bottom we think of our busyness not as a way of life we chose and are therefore responsible for but one imposed on us by forces beyond our control. In our minds, we're like an old Looney Tunes character who's walking along the street without a care in the world when suddenly an anvil falls on his head. While the cartoon anvil literally flattens Daffy Duck, ours crushes us in a different way. It's not our bodies that lie pinned beneath our busyness, it's our inner selves, those mysterious beings that live in and through our bodies, perceiving, thinking, and feeling life as it happens, moment by moment.

We tend to think of life in outward terms, as a series of events that unfold in the physical world we all inhabit, as perceived through the senses. However, we *experience* those events inwardly, in our thoughts and feelings, and it's this interior version of the world, what one leading neuroscientist has dubbed the "movie-in-the-brain," that is reality for each of us. This part of our life goes by different names: mind, spirit, soul, self, psyche, consciousness. Whatever you call it, it's this essential "you" and "me" that's squirming under the burden of too much to do and think about.

"So what?" some might say. Life has always been an exhausting grind, and dealing with it is just part of being human. And there are people who seem to enjoy being extremely busy, with never a free moment. Perhaps the rest of us should be more like them, learn to see the upside of hectic. In short, all we really need is a change in attitude.

It's tricky generalizing about something as broad and subjective as the quality of our consciousness, but there is a problem with extreme busyness that attitude alone can't fix. When it comes to creating a happy, fulfilling interior life, a "movie-in-the-brain" that makes you want to stand up and applaud, one factor matters more than any other: depth. We all know what depth is, though it's hard to pin down precisely with words. It's the quality of awareness, feeling, or understanding that comes when we truly engage with some aspect of our life experience.

It can be anything at all—a person, a place, a thing, an idea, or a sensation. Everything that happens to us all day long, every sight and sound, every personal encounter, every thought that crosses our minds is a candidate for depth. We're constantly sifting among these options, deciding where to deploy our attention. Most float around in the periphery of our thoughts and remain there, but a select few wind up in the mental spotlight. We train our perceptual and cognitive resources on one conversation, one fascinating idea, one task to the exclusion of all others. This is where depth begins.

When you're driving your car and you come to a stop sign, you perceive the sign and its meaning, and you react to it. But beyond this automatic, almost mechanical act, you don't give the sign any special thought or consideration. It doesn't enter your interior world and take up residence. Like countless other ephemeral objects of your attention, it remains on the sidelines, a bit player.

Five minutes later, you arrive home and your dog comes bounding up to greet you. You bend down and scratch her behind the ears, and she licks your face in that delirious, sloppy way of hers. As you enjoy the licks and smell her familiar doggy smell, you wonder what kind of day she had here at home while you were out in the world. You pick up a stick and throw it, and as she bounds off to retrieve it, you laugh at the eager expression on her face. Interacting with the dog floods your consciousness with thoughts and feelings. Unlike the stop sign, the moment has richness and texture that you experience in all its fullness. You are there with your dog and no place else. The experience has depth.

It might seem that this is merely a function of the time you spend with the dog: the more time you give to an experience, the deeper you go. But it isn't that simple. A glance thrown across a crowded room can have more depth than a two-hour conversation. It's ultimately not a product of time or any other quantifiable attribute. Rather, it's about the inner life that a given experience takes on—its meaning. "It all depends," William James once wrote, "on the capacity of the soul to be grasped, to have its life-currents absorbed by what is given."

We've all experienced this, and we know what it does for us. The moments we enjoy most as they unfold, and that we treasure long afterward, are the ones we experience most deeply. Depth roots us in the world, gives life substance and wholeness. It enriches our work, our relationships, everything we do. It's the essential ingredient of a good life and one of the qualities we admire most in others. Great artists, thinkers, and leaders all have an unusual capacity to be "grasped" by some idea or mission, an inner engagement that drives them to pursue a vision, undaunted by obstacles. Ludwig van Beethoven, Michelangelo, Emily Dickinson, Albert Einstein, Martin Luther King, Jr.— we call them "brilliant," as if it were pure intelligence that made them who they were. But what unites them is what they *did* with their intelligence, the depth they reached in their thinking and brought to bear in their work.

It's not only geniuses who possess this quality. There are ordinary people everywhere who, through sheer joyful engagement, seem to find depth in everything they do. This enviable talent can appear preternatural, like something you have to be born with. William James acknowledged that there are lucky individuals who are so alive to experience that they can find "inner significance" in a cloudy sky or the faces of strangers on a busy city street. He wondered if there's anything the rest of us can do to acquire this extraordinary kind of awareness. "How," he asked, "can one attain to the feeling of the vital significance of an experience, if one have it not to begin with?" James reached the same conclusion that many other philosophers down through the centuries have reached: every life has the potential to be lived deeply.

That potential is lost when your days are spread so thin, busyness itself is your true occupation. If every moment is a traffic jam, it's impossible to engage any experience with one's whole self. More and more, that's how we live. We're like so many pinballs bouncing around a world of blinking lights and buzzers. There's lots of movement and noise, but it doesn't add up to much.

Now and then it occurs to us that we could do better, reconfigure our commitments and schedules so they're not so crazy and we can breathe. But no sooner do we have this thought than we dismiss it as futile. The mad rush is the real world, we tell ourselves. We're resigned to it in the same grim way that people in repressive societies become resigned to their lack of freedom. Everyone lives like this, racing and skimming their way through their days. We didn't drop the anvil, and there's nothing we do about it except soldier on, make the best of it.

Though it is indeed the norm in our society to live this way, we're kidding ourselves when we deny responsibility for it. True, some of the activities and obligations that fill our hours aren't really a matter of choice. When the boss asks you to work overtime, you do it. When the mortgage bill is due, you sit down and pay it or else. Yet beyond these involuntary time eaters, we create a lot of our own busyness by taking on tasks that nobody requires us to do. Some of those optional pursuits are enjoyable and fulfilling, such as the hobbies and causes we care about and work hard on. And some are frivolous and pointless, such as the time we spend shopping for things we don't really need. Worthwhile or not, the point is that a great many of these busy-making activities are completely our own doing. We don't just choose them, we *pursue* them.

In the last few decades, we've found a powerful new way to pursue more busyness: digital technology. Computers and smart phones are often pitched as solutions to our stressful, overextended lives. And in many ways they do make things easier, reducing the time and trouble it takes to communicate and perform important tasks. But at the same time, they link us more tightly to all the sources of our busyness. Our screens are conduits for everything that keeps us hopping—mandatory and optional, worthwhile and silly. If you have a mobile number, an Internet browser, and an e-mail address, endless people and organizations are within your reach. And you are within theirs.

We've adopted this way of life eagerly, both as individuals and as a society. For the last decade, we've worked hard to bring digital connectedness into every available corner of existence and, once it's there, to make it ever faster and more seamless. Dial-up connections gave way to high-speed broadband, which then became wireless and mobile. And we're always upgrading, looking for higher speeds, wider coverage. Meanwhile, *within* our connected lives we're continuously expanding the degree and intensity of our ties to others. Many of us have multiple inboxes and accounts, with ever-expanding lists of contacts. We sign up for the latest social and professional networks and join subgroups and circles within those networks.

Even as the number of people we're connected to rises, so do the frequency and pace of our communications. When we were still emerging from the analog age and the technology was slower, days and weeks would go by when we didn't hear from a friend or family member. Today we're in touch by the hour, the minute. It wasn't so long ago that people who received two or three hundred e-mails a day were considered outrageously busy, figures of pity. Now they're mainstream. In terms of sheer quantity, the most connected are just a few years ahead of the rest of us. A news story about a young woman in California who racked up more than 300,000 text messages in a single month is a glimpse of where we're headed. "Sacramento Teen Says She's Popular," read the subheadline. What will be the definition of popularity a decade from now?

The goal is no longer to be "in touch" but to erase the possibility of ever being out of touch. To merge, to live simultaneously with everyone, sharing every moment, every perception, thought, and action via our screens. Even the places where we used to go to get away from the crowd and the burdens it imposes on us are now connected. The simple act of going out for a walk is completely different today from what it was fifteen years ago. Whether you're walking down a big-city street or in the woods outside a country town, if you're carrying a mobile device with you, the global crowd comes along. A walk can still be a very pleasant experience, but it's a qualitatively different experience, simply because it's busier. The air is full of people.

Someone you know has just seen a great movie. Someone else had an idle thought. There's been a suicide bombing in South Asia.

Stocks soared today. Pop star has a painful secret. Someone has a new opinion. Someone is in a taxi. Please support this worthy cause. He needs that report from you—where is it? Someone wants you to join the discussion. A manhunt is on for the killers. Try this in bed. Someone's enjoying sorbet, mmmm. Your account is now overdue. Easy chicken pot pie. Here's a brilliant analysis. Latest vids from our African safari! Someone responded to your comment. Time's running out, apply now. This is my new hair. Just heard an awesome joke. Someone is working hard on his big project. They had their baby! Click here for the latest vote count . . .

It's flying so fast, we're always playing catch-up. The deeper we get into this way of life, the more I think my friend Marie's old mantra really could serve as the go-to greeting of the digital age. How are you? Busy, very busy.

Part of the problem is that we know from experience that busyness and depth are not mutually exclusive. We've all had moments when we were busy in a good way, pivoting nimbly from task to task, giving our all to the one that was in front of us at any given moment. This is how great surgeons work, performing numerous difficult procedures in a single day but serially, so that each gets full attention in its turn. Having a lot on one's plate imposes a certain discipline. There's truth in the old saying that that more you have to do, the more you get done.

Unfortunately, digital busyness usually doesn't work like surgery. Dozens of tasks jostle and compete for our attention on the screen, and both software and hardware are designed to make it easy to hop around. So easy, it's irresistible. The cursor never rests in one place for long, and neither does the mind. We're always clicking here, there, and everywhere. Thus, although we think of our screens as productivity tools, they actually undermine the serial focus that's the essence of true productivity. And the faster and more intense our connectedness becomes, the further we move away from that ideal. Digital busyness is the enemy of depth.

Not everyone lives this way, of course. First there are millions in the United States and many more around the world who can't afford to buy these technologies and are shut out of their manifold benefits, except through the limited access afforded by public libraries and other institutions. This is a real problem that deserves more attention than it receives. Second are those who *can* afford the latest gadgets but choose instead to be lightly connected or not connected at all. But these are the exceptions that prove the rule. The trend and all the momentum are emphatically in the opposite direction. The global society to which we all belong is dramatically more connected than it was a decade ago, and becoming more so each day. This shift is affecting everyone, including those who are not fully participating in it.

This is not a small matter. It's a struggle that's taking place at the center of our lives. It's a struggle *for* the center of our lives, for control of how we think and feel. When you're scrambling all the time, that's what your inner life becomes: scrambled. Why are we doing this to ourselves? Do we really want a world in which everyone is staring at screens all the time, keeping one another busy? Is there a better way?

To answer tough questions like these, we're trained to look outward, to studies and surveys that academics, pollsters, think tanks, government agencies, and others conduct on every imaginable aspect of our lives. In fact, there's a great deal of ongoing research about connective technologies and how they're affecting individuals, families, businesses, and society at large. New findings are released all the time and reported widely in the news media, where technology is a perennially hot topic: "Americans Spend Eight Hours a Day on Screens"; "Study: U.S. Loaded with Internet Addicts"; "Texting and

Driving Worse than Drinking and Driving." We read these headlines and shake our heads, not because they're telling us something we don't know but because we know it all too well. The reality of our connected lives is all around us. What these broad findings don't tell us is how to change it.

Studies and surveys reflect what's generally true, i.e., true for most people in a given population. These general truths are supposed to help us answer the questions we have about our own particular lives. In short, they look to the crowd for understanding. On some subjects, the crowd really does have the answers. In politics, for instance, elections are decided by how most people vote. That is why, in the weeks leading up to a big election, polling data are genuinely interesting and useful. To the extent that studies shed new light on some specific aspect of *how* we are living with technology, they can be illuminating; I cite some such studies in this book. However, on the question of how to respond to the challenge of screens and their growing power over us, there is no reason to believe that what most people do and say will tell us anything useful at all. To the contrary, with screens the problem is the crowd itself and why we're drawn to it so powerfully. It's like asking a chocolate layer cake to help you think about your overeating.

Ultimately, human experience is not about what happens to most people, it's about what happens to each of us, hour by hour and moment by moment. Rather than using the general as a route to the particular, sometimes we need to take exactly the opposite approach. This is especially true when the question is the quality of our lives. In recent years, there's been a tremendous fascination with crowd thinking and behavior. The digital crowd not only has power, we're told, it also has wisdom.

Watching the crowd can certainly tell you which way popular tastes are heading and who's buying which products at any given moment. This isn't wisdom at all, however, but what's commonly known as "smarts," that canny ability to read the landscape that serves one well in stock picking, gambling, and other short-term pursuits. Every crowd is just a collection of individual selves, and to understand what's happening to those selves right now, we all have instant, no-password access to the most reliable source of all. Our own lives can teach us things that no data set ever can, if we'd just pay attention to them.

To help you think about your own connected life, I'm going to begin with two stories from mine. The first is about the urge to connect to others through screens—where does it come from, and why is it so urgent? The second is about the opposite impulse, the desire to disconnect. My experiences won't be exactly like yours. I offer these stories as illustrations of the conflicting drives so many of us are feeling lately, in our own particular ways. What we haven't figured out is how to reconcile these drives or whether they even *can* be reconciled. It's the conundrum at the heart of the digital age, and in order to solve it, we first need to see it up close, in the granular details of the everyday.

Chapter Two

HELLO, MOTHER

The Magic of Screens

I'm in my car driving to my mother's house. She lives about two hours from me and close to one of those small-city airports where it's easy to park out front, the lines are short, and the security people are friendly. When I travel for work, I try to book my flights out of that airport and I get to visit my mother on both ends of the trip. This time I'm catching an evening flight, and she's cooking dinner.

As usual, I got a late start and won't be arriving anywhere near the time she's expecting me, so I need to call her to say I'll be late. I wait for a stretch of empty highway where it feels safe to look away for a few seconds. I open my mobile phone and hit the 4 key, which is programmed with her home number.* A photo of my mother appears on the screen, a head-and-shoulders shot that I took months ago with the phone's camera. I later selected it as her ID photo, so it comes up automatically when I call her or she calls me.

I really like this image of her, and I contemplate it for a

In some states I would have been breaking the law by making a phone call while driving. I made the call in Massachusetts, where at the time that I wrote this book, it was still legal, though probably not safe.

moment before putting the phone to my ear. She's wearing a pink-and-white-striped sweater and looking up at the lens with a certain cat-that-swallowed-the-canary expression she always gets just before bursting into a laugh. She laughs a lot, so this is a characteristic look for her. In other words, the photo captures something essential about my mother.

When she answers, I tell her I'm on my way but running a little behind. She chuckles knowingly. We've had this conversation so many times, it's Kabuki now and we both know our parts. She says she'll hold dinner and why don't I call again when I'm twenty minutes away? I agree to do this and tell her I can't wait to see her. We sign off.

I take the phone from my ear, glance again at the photo, then hit "END" and watch it disappear. Driving along, I feel an unexpected surge of emotion. I'm thinking about how fun it always is to spend time with my mother, how lucky I was to be born to such a warm, companionable person. Lately I've noticed shades of her humor in my son, and I wonder now if he somehow inherited that from her. Have they isolated a gene for good-naturedness?

As the minutes pass and I drive along, these thoughts about my mother flow into new ones. In my consciousness, the smile from the photo merges with the pine woods on either side of the highway and the jazz playing on the radio, beamed down from a satellite miles above the earth. Memories rise up out of nowhere and flit around me in the car. They're not specific memories of particular events but rather scenes in which I see my mother doing normal, habitual things. In the video archive of the mind, these would be the generic clips I've filed under "Mom." There she is walking across a lawn. Sitting under a beach umbrella with a book. Talking to someone at a party. Holding her sides as she breaks up over a funny story. For a while, the car is a floating cloud of filial affection and, well, joy.

It's extraordinary, this feeling of time out of time. Everything dreary and confusing about my quotidian life has dropped away. I'm not the rushed, cornered, inadequate creature I often feel like. I'm absorbed in these memories, which seem to come from a place both beyond me and deep inside me, as if far and near, outward and inward, have come together in a new harmony.

My mother and I are no longer connected in the literal sense, as we were minutes earlier. Yet I'm feeling a connection to her that is stronger than the one we had when we were actually chatting. Even as I enjoy this, I find myself thinking about the tool that engendered it, the unprepossessing, low-end clamshell-style phone now sitting dormant in the cupholder. How did it do that?

THIS EXPERIENCE IS a microscopic example of life in the early twenty-first century, just one digital connection out of the billions that now transpire every day. However, if we step back and examine it a little more closely, there are certain basic elements that figure in just about every connection and in everyone's connected experience.

First, notice that it all began with an utterly practical need. I was running late for an appointment, and I needed to notify the person who was expecting me. In this sense, it doesn't matter that that person was my mother. I was using my mobile phone to perform a simple, utilitarian task. The call represents all the useful tasks our screens enable each day, not just in family and private life but in the working world and everywhere else.

Not so many years ago, this particular useful task would have required a lot more time and effort. I would have had to stop the car and find a landline pay phone, probably at a gas station or a highway rest stop, where it would have been necessary to pull over, park, and get out of the car. The phone, in all likelihood a grimy, graffiti-smeared affair, would have required coins or a phone card, meaning more time and bother. The whole cumbersome process would have cost me at least ten minutes, making me that much later for dinner. Instead, I hit one button, instantly reached the party I needed to reach, and accomplished my goal without losing a minute of drive time.

This was not an earthshaking achievement. The trouble and time I saved weren't valuable to anyone but me and my mother, and even then, ten minutes is hardly a big deal. Yet it was the very triviality of it, paradoxically enough, that made it meaningful. Life is full of tiny moments like this one. Need to find somebody's address? Order a pizza? Copy a colleague on that memo you wrote yesterday to the vice president for sales? Find out how you did on the math test? Pay a bill? Check the weather? Many of the tasks that we use our computers and cell phones to accomplish are mundane and, by themselves, seemingly insignificant. But taken together, they add up to something very important.

After all, we spend most of our time and energy on the practical side of life, the ceaseless flow of routine tasks performed all through the day. And there's little choice in that matter. Before you can get on to the more consequential tasks you *really* care about, you have to take care of the small stuff. If you don't pay the mortgage, you won't have a house to shelter you and your loved ones. If you don't book the tickets and renew your passport, the dream vacation will never happen. If you don't check your work inbox regularly, there goes your brilliant career. In effect, our highest goals and dreams, everything we're shooting for in life, is riding on our ability to plow through those practical to-do items as efficiently and effectively as possible.

Though ten minutes is a paltry gain, when you multiply it by the number of practical tasks performed in a typical day, the potential savings in time and energy are considerable. This is the first and most basic reason why, in the last two decades, human beings have embraced digital technologies and reorganized their lives around them. And why it makes good sense that we have done so. Computers and smart phones make it much easier to accomplish the small workaday jobs that are the foundation, the sine qua non, of our larger lives and ambitions.

Our culture reminds us every day how useful these devices are, and exhorts us to take advantage of this by making sure we are as digitally connected as current technology allows. "Get Connected!" urges the cover of *Parade* magazine, one of the more reliable windows into the mind of middle America. The cover photo shows a celebrity comic wearing his trademark wacky grin. He's surrounded by digital devices, and there's a USB plug coming out of his ear. Inside are articles explaining how digital technology "is putting politics back in your hands," "bringing people together in unexpected ways," and "can make your life easier." And there's "A Bonus Pull-out Section" about "your digital home." The emphasis is on the practical: save yourself trouble and prosper with the new connectedness.

It's a no-brainer, not just for individuals but for businesses, government agencies, and organizations of all kinds. In this highly competitive world, speed and efficiency are the name of the game. Technology is about cutting costs, expanding reach, and streamlining management, thereby improving (if all goes well) overall performance and the bottom line. Again, a brilliant way of getting the small goals accomplished in the service of much larger ones. These tools have also made it much simpler for individuals with common interests and goals to find one another and create new organizations and movements. The advent of what writer Clay Shirky calls "ridiculously easy group-forming" has brought down repressive political regimes, helped communities respond to natural disasters and terrorist attacks, and worked countless other wonders of human cooperation and problem solving.

Why do you see people urgently staring into screens everywhere you go? Pick out some nuts-and-bolts task from your own professional, community, or personal life, the equivalent of my mundane need to call Mom about dinner, and you have part of the answer.

But there's much more. In addition to helping us do the everyday work that supports our ultimate aims, screens can serve those higher ends directly. Think again about my phone call. While for practical purposes it didn't matter who I was calling, spiritually and emotionally it mattered hugely. I was calling the woman who gave birth to me, a person with whom I have a relationship unlike any other. The phone brought that person's voice and personality—and, through the photo, a sense of her physical presence—into the car with me. It gave rise to that moment of pure Mom-ness, which, though brief, was extremely valuable to me, so much so that I remembered it long afterward. That moment stayed with me because it took me out of the nitty-gritty burdens and distractions that tend to dominate my thoughts and allowed me to go deep. I appreciated in a new way my relationship with one special person, and the movie-in-the-brain that is my inner life went from a yawn to a blockbuster.

What more could you ask for? In an ideal world, our days would be full of experiences like that one, brimming over with the "vital significance" that is the essence of a good life. And those experiences don't grow just out of personal relationships and interactions. Ideally, our work should be just as significant to us, palpably, while we're working. Every moment of every day is a candidate for this depth of engagement and feeling.

Digital devices can and do make this happen. We use them to nurture relationships, to feed our emotional, social, and spiritual hungers, to think creatively and express ourselves. It's no exaggeration to say that, at their best, they produce the kinds of moments that make life rewarding and worth living. If you've ever written an e-mail straight from the heart, watched a video that you couldn't stop thinking about, or read an online essay that changed how you think about the world, you know this is true.

In this particular case, it all happened because of a simple phone call. But notice that it happened *after* what we typically think of as the connection, the call itself, was over. There was a gap between the practical task and the deeper experience that followed. If that gap had not been there, would I have reaped the same benefits? Doubtful. If I'd kept on using the phone for other tasks, there wouldn't have been time or space in my thoughts for the moment to unfold as it did. The same goes for any screen task with the potential for deeper impact and value, and many do have that potential—it *could* happen, but only if you give it room. We don't know about those lost opportunities, of course, because they never see the light of day. But I think we miss them, nonetheless, whenever it occurs to us that life isn't quite hanging together, isn't adding up to what it might be. It's all those unrealized epiphanies, insights, and joys—journeys the mind and heart never get to take.

If you're sitting in the office zipping from e-mail to e-mail to text to Web page to buzzing mobile and back again—that is, doing the usual digital dance—you're likely losing all kinds of opportunities to reach the depth I'm talking about. An e-mail from a client requesting an innovative improvement in the product you sell might inspire you to draw up a brief sketch of how to make it happen. Heck, you might be motivated to go home and do it yourself and perhaps start your own company selling this superior product, taking on your current employer and shaking up the whole market. It could change your life. But if you never pause to allow that thought to blossom and instead move on to the next tiny screen task and then the next and the next, guess what? No new life for you.

The gap in time between my call to Mom and the "payoff" it yielded is tremendously significant. It's the essential link between the utilitarian side of the digital experience and the "vital significance" side. And it's a link that's completely over-looked in current thinking about technology, with its unexam-ined faith in nonstop connectedness.

This is not to say the technology industry ignores the deeper potential of these gadgets. To the contrary, it adver-tises it, literally, because it's crucial to the appeal of the prod-ucts. If they did only menial jobs, we would view them roughly the way we view our vacuum cleaners. Instead, we think of them as friends, muses, passports to higher realms, and that's how they're marketed. A few years ago, there was an arresting television commercial for what was then the most fashionable mobile device, Apple's first iPhone. A gorgeous young woman was shown standing by herself against a simple black backdrop, holding the sleek wedge in her hand. A dancer with the New York City Ballet, she talked about how she used it to mobile-blog about her art from backstage during performances. "It's multitasking," she said perkily. "It's important. Even for ballet dancers."

Now, Apple could have cast any attractive, articulate multi-tasker in the commercial. But the ballerina spun the message in a very particular way. This tool, she suggested, isn't just for utilitarian drudgery. It's for the artist and the spiritual seeker in each of us. For the *soul*.

The creative potential of digital tools is very real, and it's manifest in the exuberant, richly inventive culture that has grown up online in a relatively short time, perhaps best exemplified by all the playful new additions to our language, from Googling to tweets. This imaginative and inspirational dimension of screen life is as relevant to organizations as to individuals. Whether you're running a small business, a university, a hospital, or a global conglomerate, there's nothing more valuable than an employee with a fertile, creative mind. Witness the countless business management articles and books about how to "think outside the box," make conceptual leaps, tap the right side of the brain, discover hidden strengths. Under the best circumstances, digital screens can help us do all these things. That is the other reason, beyond pure efficiency, why they are essential tools of every modern organization. They bring out the inner ballerina in us all.

But here's the question: if the ballerina is using her smart phone for serious multitasking—not just doing the blog but jumping among many other tasks at the same time, racing from this to that and back to this again—is she really tapping her inner muse, as the ad implies? Is she using the tool to optimal effect?

My phone call to Mom demonstrates the two essential benefits digital connectedness confers: we can get everyday jobs done more easily *and* nurture our minds, hearts, and souls, all with a little gizmo that fits into our pockets. It's a combination that doesn't come along very often, and it explains why screens have taken the world by storm, inspiring the kind of intense devotion and popular mania typically associated with political movements and religious crusades. For some people, digital technology isn't just a new kind of tool, it's a revolutionary creed to believe in and live for, a movement that's transforming and perfecting life on Earth. The Answer.

This view has been on especially prominent display in the news media, where major political, social, and cultural events are increasingly seen through the prism of technology. When dramatic news breaks out somewhere in the world, whether it's a terrorist bombing in London or a prodemocracy uprising in Iran, often it's not the substance of the events or the stories of leading figures that garner the most avid coverage; it's the role played by technology. A freedom march is one thing, but a freedom march planned and executed via digital gadgets—that's news.

When the news is explicitly about these devices, the zeal hits a fever pitch. In the summer of 2008, for instance, a new, improved version of the handheld endorsed by the ballerina, Apple's iPhone 3G, came to market. The company's CEO unveiled it at a press conference that, in one tech journalist's description, was more like a political rally or evangelical revival meeting:

> I've been to enough Steve Jobs keynotes now to know that the man is able to take a crowd and bend it to his will. Every time, I've been a willing subject—sometimes (but not every time) to find myself in a hangover-like state a day later when I try to remember exactly why I thought that whatever he was pitching would change my life forever. Steve Jobs is masterful and charismatic when he's on stage and all eyes are on him. And when, like yesterday, the crowd is carefully packed with a throng of Apple developers cheering him on, the press in attendance can easily get caught up in the hype.

Indeed, much of the resulting coverage had a true-believer ring to it. Here was a device that could do anything, from the utterly mundane ("Where to Eat? Ask Your iPhone" headlined one of the nation's leading newspapers) to the truly heroic

("Can the iPhone Really Save America?" asked another, only half jokingly). Here again, journalists were simply reflecting the public obsession with these devices, rooted in widespread personal experience. It's as if, having seen firsthand the potential of digital devices, we really believe that Nirvana is just an upgrade away. Consumers desperate to be among the first to own the new model formed long lines outside stores, in some cases even sleeping overnight on the sidewalk to secure the choicest spots. In California, some people were standing in line not because they wanted to buy the device but to participate in what one news report described as a "tribal experience." As one of the line standers put it, "I'd like to be part of the magic."

This magic is why we have migrated so much of our lives to the digital sphere. It's why we go through the day basically tethered to our screens. It's why these devices have proliferated at such an astonishing pace. The total number of mobile phones in the world went from about 500 million at the beginning of this century to approaching 5 billion today.

But there's a missing piece: the real magic of these tools, the catalyst that transforms them from utilitarian devices into instruments of creativity, depth, and transcendence, lies in the gap that occurred between my phone call to Mom and the powerful experience that followed. That gap was the linchpin, the catalyst. It allowed me to take a run-of-the-mill outward experience and go inward. It's the same for every kind of digital task. If you pile them on so fast that screen life becomes a blur and there are no gaps in your connectedness, you never get to that place where the most valuable benefits are. We're eliminating the gaps, when we should be creating them.

A few years ago, a leading high-tech research firm surveyed people in seventeen countries, including the United States, China, India, Russia, Germany, and Japan, with the goal of "quantifying the state of today's connectedness." The subjects of the study were quizzed about how often they connect digitally, where they connect, the devices they use, and so on. Based on the answers, the authors created a taxonomy of human connectedness comprising four different types: Hyperconnected, Increasingly Connected, Passive Online, and Barebones Users.

The study focused on the Hyperconnected, defined as "those who have fully embraced the brave new world, with more devices per capita . . . and more intense use of new communications applications."

In 2008, just 16 percent of the world's working population qualified as Hyperconnected, but the study predicted that 40 percent of us would soon meet the criteria. Bear in mind that this white paper was sponsored by a large high-tech company and had a distinctly boosterish point of view, as suggested by its title, "The Hyperconnected: Here They Come!" But let's assume for the moment that it's basically accurate and this is where the world is headed. What would it really mean to live this way? At the time of the study, the average Hyperconnected person was using at least seven different digital devices and nine different applications, in order to stay as screen-connected as possible at all times, including "on vacation, in restaurants, from bed, and even in places of worship." As this group grows larger, it predicted, there will be a "profusion" of new devices.

Of course, even as the gadgets are multiplying, they're also converging. Smart phone–like devices that will allow us to conduct our entire connected lives through one screen are currently thought to represent the future. Ultimately, it doesn't much matter how many or how few different devices we use to connect. The question is whether the hyperconnected life is taking us where we want to go.

Collectively, we seem to have decided it is. In societies around the world, obtaining the very latest, fastest, most extreme version of connectedness isn't just a core individual goal, it's a national ambition. Countries are engaged in a global race to become the most connected society on earth. Rather than sleeping on the sidewalk, they're setting national policies and spending money in pursuit of this goal. The Organisation for Economic Co-operation and Development, a Paris-based group of about thirty wealthy industrialized countries including the United States, tracks the spread of digital technology and regularly ranks countries by broadband Internet "penetration," or the percentage of the population with broadband service. These rankings are closely watched by global business and political leaders as a key marker of national status. In recent years, a handful of Asian and northern European countries have dominated the top rungs of the OECD list, while the United States has languished in the middle of the pack. In short, our relative connectedness has been abysmal.

Members of Congress decry the situation and call for a national effort to end this "broadband divide." Newspapers publish concerned editorials, think tanks issue policy papers. It's all strikingly reminiscent of the panic that followed the Soviet launch of the Sputnik satellite in the late 1950s, when the United States suddenly realized it was behind in the space race. But rather than shooting for the moon, we're after more data per second. In his 2008 campaign for president, Barack Obama said the United States should "lead the world in broadband penetration." If elected, he promised, he would do whatever was necessary to make that happen. Soon after winning election, he repeated the promise, calling it "unacceptable that the United States ranks fifteenth in the world in broadband adoption" and noting that America's future competitiveness was at stake.

There's no question that this is true, if competitiveness is measured purely in terms of who is most connected, which is what the national rankings reflect. It goes without saying that the United States and every other developed country should work hard to provide digital access to all those who lack it. But this worthy goal of bringing broadband to all citizens is not the same thing as devoting yourself to being the most connected society on earth, which at a certain point could actually be bad for competitiveness. As the top countries close in on 100 percent broadband connectedness, the question will inevitably become whose hyperconnectedness is more hyper. One reason South Korea has long ranked among the most connected countries on earth is that it's more obsessed with online gaming than any other culture. Screen games are fun and, at their best, educational. But spending huge chunks of one's day gaming is demonstrably *not* good for personal productivity.

Aren't a society's competitiveness and its prospects for a better future rooted in more than sheer technology? Isn't how well we use the devices just as crucial as how fast they are? Will pursuing more and more digital connectedness make us smarter and more creative? Will it help us understand one another better? When we're all hyperconnected, will our families and communities be stronger? Will we build better organizations and lead more prosperous lives?

Most important, can we accomplish any of these lofty goals if we continue devoting all our energy to eliminating the very thing we need most to achieve them in the first place—some space between tasks, respites, stopping places for the mind?

We don't ask these questions because they're philosophical and, unlike technology, which is concrete and quantifiable, philosophy seems abstract and squishy. So we avoid them, focusing instead on the tools themselves, breathlessly trying to keep up with the hot new devices and the latest trends. This is shortsighted since, in the end, it's the philosophical questions that really matter. Someday, it will be hard to remember why we were once so fired up about 3G connectivity and the wonders of mobile broadband. Seamless, lightning-fast connectedness will be a given everywhere on Earth, and today's gadgets will be quaint museum pieces. At that point, all we'll care about is what kind of life these devices have created for us. And if it isn't a good life, we'll wonder what we did wrong.

Right now, in these early years of the digital era, without even realizing it, we're living by a very particular philosophy of technology. It can summarized in a sentence:

It's good to be connected, and it's bad to be disconnected.

This is a simple idea but one with enormous implications. Once you assume that it's a good thing to be connected through digital networks and a bad thing to be disconnected from them, it becomes very clear how to organize your screen time and, indeed, every waking hour. If digital connectedness is intrinsically good, it follows that one should try as hard as possible to stay connected at all times or, to put it another way, avoid being disconnected. Thus, our philosophy has two corollaries:

First corollary: The more you connect, the better off you are.

Second corollary: The more you disconnect, the worse off you are.

Together, these two propositions prescribe exactly how to manage one's digital existence. You can't be too connected, they say, so we should seek at all times to maximize our time with screens and minimize our time away. And this is just how many of us are living today. We are *digital maximalists*.

Having just explored the broad range of very real benefits

that these devices offer, it's easy to see why we've embraced this philosophy and the way of life it produces. Those benefits are manifest all around us, so manifest that for many years now it hasn't seemed necessary even to wonder about the philosophy behind it or question its tenets. Connecting enhances life on so many levels, it's common sense to conclude that one should be as connected as possible all the times. Digital maximalism is clearly a superior way of living.

Except when it isn't.

Chapter Three

GONE OVERBOARD

Falling Out with the Connected Life

It's a sunny morning in late spring and I'm on the water in an old boat. Several years ago, my family and I moved from the suburbs of Washington, D.C., to a small town on Cape Cod. We'd grown tired of the city, particularly the brutal traffic that ate up so much of our time. It was a very digital-age move, an example of the endless possibilities these technologies have opened up for working at a distance in time and space from the traditional workplace. This ability to "time-shift" and "location-shift" holds the promise of a kind of liberation, a chance to own one's life more fully. Since my wife and I are both writers, as long as we have our screens and a good connection, we can stay in touch with all the people and sources of information we need to get our work done. Friends, too, are now as reachable as the nearest digital gadget. So we decided to try living in a different kind of place. We moved to the outer reaches of the Cape, which we knew from summer vacations and had dreamed about as a place to live full-time and raise our son, who was then seven years old.

In our new hometown, people spend a lot of time in boats, and we wanted to join in. Good-bye urban gridlock, hello open water. We started watching Craigslist for used motorboats. It took many months to find one we liked and could afford. It was more than twenty years old and pretty dilapidated, but it had character and we couldn't wait to start tooling around in it. We named it *What Larks!* after a line from *Great Expectations*, the Charles Dickens novel about the coming of age of a boy named Pip. When Pip's brother-in-law, the kindly blacksmith Joe Gargery, talks about the good times the two of them will have together someday, that's what he always says, "What larks!" Life should really be full of larks, and we hoped to have many in our boat.

So I'm out here getting it ready for its first season. I've just started the engine, and now I need to back away from the dock, where the boat was tied up for a few days for repairs. As a novice boater, I'm a little nervous about this. There are several other boats moored in the area I'm backing into, and my task is to maneuver between them while swinging my boat around. It goes smoothly at first, and I'm almost home free when I notice I'm passing dangerously close to another boat's mooring line. I hear my propeller struggle, then seize up.

I turn off the engine and peer over the stern to examine the situation. The line is wrapped tightly around the propeller several times. But if I reach down, bracing myself on the engine with one hand, I'm almost sure I can use my free hand to unwind it. I attempt to do this, but it's quite a reach and I'm leaning farther out than planned, so much farther I'm worried I might be in danger of . . . *Yiiiiiiiiikkkes!* I tumble into the water headfirst, fully clothed.

On surfacing, the first thing I do is look around to see if there were any witnesses to my ignominy, which I'm sure looked like a clip from one of those world's-funniest-videos shows. Happily, there's nobody in sight. The second thing I do is feel underwater for my credit card wallet and mobile phone, which I always keep together in my left front pocket. Both are still there, phew.

Wait, they're both there? No! The wallet will be fine, but my phone is drowning. Panicked, I pull it out and throw it into the boat, then quickly untangle the propeller. I clamber back in and pick up the phone. It's buzzing in a strange, halfhearted way that it's never buzzed before, the digital version of a last gasp. I frantically press every button on the keypad, but nothing happens. The screen is completely blank. After about a minute, the buzzing stops. It's dead.

I've owned many mobile phones in my life, but I've never killed one before, and I'm mad at myself. I think of all the names, numbers, and e-mail addresses I have stored on it, the dozens of photos that I never backed up. What a pain it will be to get a new phone, choose a model, decide on the length and other details of my new contract (this time I *will* get the insurance), reload all my contacts, who knows what else. Sometimes when I'm down in the weeds of my connected life, managing my relationship with all these distant technology companies and service providers, I think of a phrase Thomas Jefferson used in a completely different context: "entangling alliances." Right now I have a yen to disentangle myself and follow a more isolationist path.

Then reality sets in. Fact is, I need those entangling alliances. They keep me in touch with everyone I care about, not to mention my sources of income. And though it's true that my screens have added a whole new layer of complications to my life, in other, arguably more significant ways, they've simplified and improved it. Without them, I wouldn't be living in this far-flung place, where I feel more "connected" to my own life and those around me than I've ever felt anywhere else. If the phone is my prison keeper, it's also my liberator, and I'll need to start looking for a new one immediately.

There may be a day or two when I'll have to be completely phoneless. What a disaster.

Minutes later, heading back across the cove, I notice something funny. It's not anything I can see or hear. It's an inner sensation, a subtle awareness. *I'm completely unreachable.* Friends and family can't reach me. Colleagues and contacts from my work life can't reach me. Nobody anywhere on the planet can reach me right now, nor can I reach them. They're out there in the great beyond, and, short of Jedi-like telepathy, there's no way of bridging the distance between us. Just minutes ago, I was embarrassed and angry at myself for drowning my phone. Now that it's gone and connecting is no longer an option, I like what's happening.

Before I went overboard, I was alone in the boat, in the classic sense of alone—there was nobody physically with me. But because I had a connective device in my pocket, in another sense I wasn't alone at all. Everyone in my life was just a few button taps away. Now I'm alone in a whole new way. It's a state I used to know very well. I remember walking around my college campus in the early 1980s, on my own in the world for the first time. This was pre–cell phone era, so when I was out in public like that, I had no easy way of communicating with most of the human race. It was a bit lonely being away from my parents and all the other people I'd always been dependent on for support and companionship. But it was also exhilarating. Here I was, a full-fledged person finally at the controls of my own life. I had some doubts that I was ready, but that was part of the thrill.

Years earlier, as an adolescent in the throes of my first existential crisis, I'd read a self-help book called *How to Be Your Own Best Friend*, a bestseller of the 1970s, now sadly almost forgotten. Written by a pair of married psychoanalysts named Mildred Newman and Bernard Berkowitz in a spare, Zen-like style, it consisted of short philosophical questions ("Why are so many people dissatisfied in so many ways?") and answers. The book's basic thesis was that in order to find peace and contentment, we must accept our fundamental separateness from others. Happiness is about knowing how to enjoy one's own company:

> Someone who cannot tolerate aloneness is someone who doesn't know he's grown up. It takes courage to let go of that fantasy of childhood safety. The world may never seem so certain again, but what fresh air we breathe when we take possession of our own separateness, our own integrity! That's when our adult life really begins.

This was a revelation to me. When I had thought of aloneness at all, which wasn't often, it was as a negative, an absence of something intrinsically good: the company of others. My own emotional experience painfully confirmed this impression. Is there anyone lonelier than a gangly thirteen-year-old with braces and thick eyeglasses? It had never occurred to me that aloneness could be a fruitful, let alone ecstatic, experience. Or that accepting and exploring my separateness might be the way out of misery and into maturity.

You never know where you're going to find wisdom. This simple thought stayed with me, resonating as forcefully as anything I gleaned from the great books I later read in high school. Indeed, I saw it echoed over and over in the fictional characters who populate our best stories, people struggling to come to terms with their essential isolation from others and thus with themselves. Odysseus, Don Quixote, King Lear, Ishmael, Hester Prynne, Huck Finn, Leopold Bloom, Holden Caulfield—it's their journeys as individuals, into their individuality, that draws us back to their stories again and again. Because they're our story, too.

The twentieth-century philosopher Paul Tillich once wrote that the word "loneliness" exists to express "the pain of being alone," while "solitude" expresses "the glory of being alone." I was experiencing both in those college days, but it's the glory that I remember most today. The older I got, the more I saw how crucial maintaining some degree of separateness was to my own inner tranquillity and, at the same time, how hard it was to achieve. Society is constantly throwing up obstacles, telling us that we're worthless without the crowd, that everything is riding on its approval.

In a country built on ideals of individual freedom and autonomy, one might think such messages wouldn't get much traction. But freedom can be a heavy burden, and in a certain sense, the more we're responsible for managing our own destinies, the more appealing conformity becomes. Recognizing this, marketers have learned to sell products in a way that makes us *feel* like bold individualists, even as we're joining the herd. Advertisements pitch everything from cars to cola as instruments of self-expression and liberation, though they're really the opposite. Be a rebel, wear the shoes everyone else is wearing.

I still struggle to ignore these messages. But when I succeed at standing apart, the payoff is enormous, and not just in a selfish way. The best kind of aloneness is expansive and generous. To enjoy your own company is to be at ease not just with yourself but with everyone and everything in the universe. When you're inwardly content, you don't need others to prop you up, so you can think about them more freely and generously. Paradoxically enough, separation is the way to empathy. In solitude we meet not just ourselves but all other selves, and it turns out we hardly knew them.

Social separateness was still plentiful in the late 1980s, when I was out of school living on my own for the first time in a big city. This was the dawn of the digital era, when personal computers were increasingly common and e-mail was first catching on. Cell phones were still rare, however, as were truly portable laptops. So when you were out in public, you were still basically disconnected. Walking the city in those days, I was both surrounded by others and utterly alone, and it was this solitude within the crowd that made city life magical. It's what E. B. White was talking about when he observed that New York City "blends the gift of privacy with the excitement of participation . . . insulating the individual (if he wants it, and almost everybody wants or needs it) against all enormous and violent and wonderful events that are taking place every minute."

Written in 1948, that line reads today like an inscription on an ancient tomb. The old unreachability, the effortless kind you could experience even in the midst of a metropolis, has vanished. Right around the start of this century, the ideal of a life blending "privacy" and "participation" was thrown out, replaced by an idealization of maximum connectedness. The first corollary, "The more you connect, the better off you are," took hold all through society, for the reasons we've already seen. Digital connectedness was a deeply compelling force, one that served manifold human purposes and needs. And right on its heels was the second corollary: "The more you disconnect, the worse off you are." Perhaps there had been a time when most everyone wanted or needed insulation from the crowd, but now they didn't seem to need it anymore. In a society built on the maximalist ideal, to be disconnected was to be out of it, cut off, in a bad place.

We never sat down and consciously decided that this was the code we would live by. There was no discussion, no referendum or show of hands. It just sort of happened, as if by tacit agreement or silent oath. *From now on, I will strive to be as connected as possible at all times.* Like everyone else, I signed right on. I've spent most of the last decade within arm's length of a computer or my phone, usually both. When I was away from technology or when I just couldn't find a signal, I perceived it as a problem. If a hotel didn't have broadband in the room, I got irritated and complained. When I found myself in a region without cell phone coverage, I felt my provider had let me down. Staying with cousins for the holidays in a house without a wireless router, and thus no Internet connection for my laptop, I would go into the backyard or sit in the car on the street and try to pick up a neighbor's signal. Not once or twice a day, but many times. How else was I supposed to know what was going on in my life?

Of course, as wireless technology improved and spread, these frustrations diminished. By the middle of the decade, it was much easier to find a reliable connection. Laptops got smaller, and cell phones acquired Internet browsers, making web access as portable as a wallet. Involuntary disconnectedness was increasingly rare. We started to view our high-speed connections the way we view electricity and running water, as a given of everyday life.

It was exactly at this point that I really started to think about my own connectedness for the first time. Given my maximalist tendencies, I should have been delighted to see digital connectivity spreading far and wide. Wasn't this what I'd wished for? No more of those irritating moments of isolation.

But here's the weird thing: I started *missing* them. It wasn't the annoyance and frustration that I wanted back—I'm no masochist. It was the state of mind that I'd found myself in *after* I couldn't get a connection and gave up. According to the second corollary, in these disconnected times, I should have felt a deterioration in the quality of my life. The more you disconnect, the worse off you are, right? Once I accepted my fate, however, I'd experience a slow but steady improvement in my overall mood and attitude. I wasn't completely conscious of this effect at the time, but it was definitely stored someplace on my inner hard drive. There I was with no inbox to check, nothing to click on or respond to. No demands, requests, or options. No headlines to scan or orders to place. No crowd to keep me busy. With all of that out of reach, my consciousness had no choice but to settle down into the physical place where I happened to be and make the best of it.

At those times, I'd been a castaway washed up on a desert island, a digital Robinson Crusoe. And in classic castaway fashion, now that I was rescued, I saw in retrospect that there had been something very special about my island. *Life was different* in disconnected mode. The easier it became to stay connected, the more I thought about this other, different way of being and started longing for it.

I first noticed it on airplanes. Cell phone use has long been prohibited on commercial flights, and at that time there was still no airborne Internet service. Boarding one of those disconnected flights was like passing through a wormhole into another dimension where time moved differently. Buckling into my seat, I felt my mind relax as I was liberated from a burden I didn't even know I'd been carrying. It was the burden of my busy, connected life. The burden of always knowing that everyone everywhere is just a few clicks away.

The limitlessness of digital life is thrilling, but it's also unsettling, in two important ways. First, the hours we spend flitting constantly among tasks train us to treat our time and our attention as infinitely divisible commodities. On a screen, it's easy to jam more busyness into each moment, so that is exactly what we do. Eventually the mind falls into a mode of thinking, a kind of nervous rhythm that's inherently about finding new stimuli, new jobs to perform. This carries over into the rest of our lives; even when we're away from screens, it's hard for our minds to stop clicking around and come to rest.

At some point, I noticed that it had become hard for me to stay focused on a single task of any kind, mental or physical, without adding new ones. While brushing my teeth, I would wander out of the bathroom in search of something else to do at the same time. I'd be organizing my sock drawer with one hand while trying to reach my wisdom teeth with the other, and even then I could feel myself craving still another job. The digital consciousness can't tolerate three minutes of pure focus.

The second unsettling aspect of this is philosophical. The more we connect, the more our thoughts lean outward. There's a preoccupation with what's going on "out there" in the bustling otherworld, rather than "in here" with yourself and those right around you. What was once exterior and faraway is now easily accessible, and this carries a sense of obligation or duty. When you can reach out and touch the whole world, a part of you guiltily feels you *should* be reaching out. Who's waiting to hear from me? Is the boss wondering why I haven't responded?

In addition, outwardness offers something more potent than mere duty: self-affirmation, demonstrable evidence of one's existence and impact on the world. In less connected times, human beings were forced to shape their own interior sense of identity and worth—to become self-sufficient. By virtue of its interactivity, the digital medium is a source of constant confirmation that, yes, you do indeed exist and matter. However, the external validation provided by incoming messages and the number of times one's name appears in search results is not as trustworthy or stable as the kind that comes from inside. Thus we're forced to go back again and again for verification. Who dropped my name? Who's read my latest post? Are there any comments on my comments? Who's paying attention to me now?

Up in an airplane without wireless, all of that receded. The world of endless potential tasks was gone, and so too was that feeling of compulsory, needy outwardness. What was special about those flights was the very thing I had long tried to avoid, involuntary disconnectedness. Though air travel is miserable in so many ways—and at six foot five, I'm no fan of the Torquemada-inspired seats of economy class—I actually started looking forward to it. Here was a rare respite from my connected life. Existence was reined in, reduced to just me and my immediate surroundings, the other passengers, the cup of tea on the tray table, the words on my notebook screen. I got some of my best thinking and writing done on those flights. And down on the toolbar at the bottom of the screen, was a constant reminder of why: the red X over the wireless icon, for no signal.

Now, PLOWING THROUGH the water with a lifeless cell phone in my pocket, I'm having the same sensation. It's bracing being out here all by myself, with nothing to distract me from the task at hand. In fact, there *is* no task at hand other than getting back to my mooring. Not only can nobody reach me, but, just as amazing, I can't click a few buttons and create busywork for myself. If my phone were working right now, I'd be on it with my wife, Martha, saying I'd be home in twenty minutes, though she really doesn't need to know that. I'd tell her about how I'd fallen in the drink, and we'd have a laugh about it.

After years of being so connected, I'm used to sharing all thoughts and experiences impulsively, in the moment, with everyone and anyone who comes to mind. Why *not*, when they're all around you? I've forgotten that some information is like wine: it gets better if you let it rest for a while.

I'm steering with one finger, watching the seabirds dive for breakfast. *Oh, what a beautiful morning, oh, what a beautiful day.* Okay, I'm not actually singing, but I could be. Could this sudden ebullience be all about a dead cell phone? No, I'm happy first and foremost because I'm out on the water on a nice spring day. But there's a special quality to this mood, a looseness in the way thoughts and feelings come and go, and even in how my body feels. It's a *happier* kind of happiness that reminds me of those early tastes of independence in the old predigital world. The little self-help book was right—*what fresh air we breathe when we take possession of our own separateness, our own integrity!*

LIKE MANY OTHERS, I'd been dutifully toeing the line, allowing digital connectedness to reshape my life without asking if it was the kind of life I really wanted. The more connected I was, the busier I became attending to all the people, information, and tasks that the devices bring within our reach. And this had two distinctly negative effects. First, as the gaps between my digital tasks disappeared, so did the opportunities for depth. Screen life became more rushed and superficial, a nonstop mental traffic jam. Second, because I was spending so much time in the digital sphere, I was less able to enjoy my own company and the places and people right around me.

The same tool that added depth to my experience was taking it away. It was only when the tool was rendered useless that I felt the balance shifting back. When my phone died, a space opened up between me and the rest of the world, and in that space my mind was able to settle down. It was an accidental

version of the place I went to after calling my mother and a reminder of how important that place is. I was myself again that
morning, free in a way I seldom felt free anymore. What larks!

Yet the opposite message was coming from all directions:
Connect! Connect! A revolution was under way and people
were sleeping on sidewalks to be in the vanguard. Standing at a
crosswalk in midtown Manhattan one day waiting for the light
to change, I realized that the eight or ten other pedestrians
standing around me were *all* staring into screens. Here they
were in the heart of one of the greatest cities in the history of
civilization, surrounded by a rich array of sights, sounds, and
faces, and they were running away from it all, blocking it out.

When a crowd adopts a point of view en masse, all critical thinking effectively stops. The maximalist dogma is particularly difficult to challenge because it's all about joining the
crowd, so it's self-reinforcing. There was an inexorability to it,
a sense that if you didn't hop on the digital bandwagon and stay
there, you'd be left behind. The industry study I mentioned
earlier, which predicted a vast migration of the human species
to hyperconnected living, included this stern warning to the
global business community: "Enterprises will either manage
this migration or get trampled." Who wants to be trampled?
Besides, the technologies *are* remarkable. To have doubts felt
retrograde, like throwing in one's lot with the technology pessimists, casting a vote against the future.

Our own perceptions and feelings are rarely as peculiar as
we think. The lonely thought you had at 3 A.M. turns out to
be everyone else's lonely thought, you just didn't know it at
the time. Once I started to question my own maximalist tendencies, I began seeing evidence that I wasn't alone. In the
news outlets I habitually follow, the same ones that cheered
the revolution and promoted digital devices as saviors, stories about the burdens of overconnectedness appeared with

increasing frequency. They weren't bannered across the top of the front page or leading the newscast. You had to look for them in the back pages, scroll down, or wait for the second half of the show. Typically, there would be some new study or survey indicating that screen life was taking a previously unrecognized toll. Some of these reports were intriguing enough that I clicked around and found the source material, which tended to be sketchy and inconclusive. Still, the point was that others were noticing what I was noticing.

Problems were turning up in three different but overlapping places: (1) in the interior lives of individuals, where experts were describing psychological and emotional disturbances far more serious than what I'd experienced; (2) in family and personal relationships, where screen time has been replacing face time; and (3) in businesses and other organizations, where distracted workers are hurting the bottom line. Let's take them one at a time.

From the earliest days of computers, there have been worries about the effects these technologies would have on the human mind. Back in the early 1970s, the futurist Alvin Toffler coined the term "information overload" to capture what he believed would happen to the human consciousness as connective technologies brought the world to our mental doorstep. In the last decade, the phrase has gained new currency, mainly through media reports about novel psychological conditions and behaviors that some experts attribute to digital overload. They include attention deficit trait, a malady related to the like-named bane of modern childhood. According to Edward Hallowell, the psychiatrist who first described it, ADT is "like a traffic jam in your mind." Symptoms include "distractability, restlessness, a sense of 'gotta go, gotta rush, gotta run around' and impulsive decision-making, because you have so many things to do."

Many other conditions have been linked to overload, including continuous partial attention, defined as the state of mind in which "most of one's attention is on a primary task, but where one is also monitoring several background tasks just in case something more important or interesting comes up." E-mail apnea, meanwhile, is "a form of shallow breathing while checking email that, in some extreme cases, leads to an increase in stress-related disease." There's also Internet addiction disorder and, at the comic end of the spectrum, nomophobia, "the fear of being out of mobile phone contact."

New, snappily named disorders are always suspicious, crafted as they often are to pull in the very media coverage in which we learn about them. For our purposes, whether they really exist as discrete phenomena is beside the point. What the media serve up each day, more than anything, is an X-ray of the collective consciousness, which is just the sum of our individual hopes and fears. When crime rates are a worry, the headlines are full of serial killers. When global climate change first entered public awareness, every big storm was a symptom. Likewise, this rash of alleged digital neuroses reflects the concern just about everyone now feels about the relentless pull of the screen. Putting a scientific-sounding name on it, however contrived, is a way of feeling in control. As with crime and climate change, this doesn't mean the underlying problem is illusory. Nomophobia may sound funny, but the challenge of the new busyness couldn't be more real.

THE INTERIOR STRUGGLE is having a dramatic impact in our personal and family relationships. If we've learned anything in the last decade about technology and human interaction, it's that as screen time rises, direct human-to-human interaction falls off proportionally. We encounter this truth every day in the small moments when our relatedness to others is interrupted and fractured by technology. The conversation broken off by another person's telephone ringing. The voice that trails away as eyes and brain tunnel into a screen.

It's annoying when you're the victim, but then, don't you do the same thing yourself? You're in a real place with someone who means a great deal to you, say, having lunch with a close friend or colleague or reading a book to a child. To all appearances, you're present and fully engaged. But your attention is provisional, awaiting the next summons from beyond. A faint vibration or beep is all it takes, and off you go.

I'd seen this phenomenon so often in my own family life, I'd given it a name—the Vanishing Family Trick. We're all together in the living room after dinner, the three of us plus two cats and a dog, enjoying one another's company. Our house is very old and snug, and the living room, a former stable with age-darkened paneling and exposed beams, is ideal for this kind of togetherness. In the winter we move the furniture closer to the fireplace, so it's even cozier. It's a natural place to hang out.

Here's what happens next: Somebody excuses themselves for a bathroom visit or a glass of water and doesn't return. Five minutes later, another of us exits on a similarly mundane excuse along the lines of "I have to check something." The third, now alone, soon follows, leaving just the animals, who, if they can think about such things, must be wondering what suddenly became of a splendid gathering that had barely gotten under way. Where have all the humans gone?

To their screens, of course. Where they always go these days. The digital crowd has a way of elbowing its way into everything, to the point where a family can't sit in a room together for half an hour without somebody, or everybody, peeling off.

What's lost in the process is so valuable, it can't be quantified. Isn't this what we live for, when you come down to it, time spent with other people, those moments that can't be translated into ones and zeros and replicated on a screen? Obviously, relationships are about more than being with others in the literal, bodily sense. They can be maintained and nurtured across great distances using all kinds of connective tools. For centuries letters served this function beautifully, allowing people to conduct elaborate, long-running dialogues that could be more intimate and affecting than in-person conversation.

E-mail plays an analogous role today, though we put less thought and care into e-mail messages than our ancestors often put into their letters. In e-mail programs the button to start a new message reads "Compose," connoting a level of artistry that my e-mails don't deserve. I slam them out, one after another, rarely pausing to consider how well they read or even to correct typos. I read most of the e-mails I receive in the same way. The point is almost *not* to be thoughtful, not to pause and reflect. To eliminate the gaps.

The rushed, careless quality of screen communication is of a piece with the discounting of physical togetherness. When everyone is endlessly available, all forms of human contact begin to seem less special and significant. Little by little, companionship itself becomes a commodity, cheap, easily taken for granted. A person is just another person, and there are so many of those, blah, blah, blah. Why not flee the few of the living room for the many of the screen, where all relationships are flattened into one user-friendly mosaic, a human collage that's endlessly clickable and never demands your full attention?

Somewhere inside, we all know this isn't the path to happiness. My most cherished childhood memories, the ones that made me who I am and sustain me today, are about moments when a parent, grandparent, or somebody else I cared about put everything and everyone else aside to be with *me alone*, to enter my little world and let me enter theirs. In the old Doors song "Break on Through (to the Other Side)," there's a line about finding a "country in your eyes." We weren't visiting one another's countries much anymore—they were becoming foreign lands. As I watched the Vanishing Family Trick unfold and played my own part in it, I sometimes felt as if love itself, or the acts of heart and mind that constitute love, were being leached out of the house by our screens.

It's been happening for a long time now to families everywhere, and nobody seems to know how to stop it. Several years ago, *Time* magazine ran a cover story about children and technology that opened with this slice of life:

It's 9:30 p.m., and Stephen and Georgina Cox know exactly where their children are. Well, their bodies, at least. Piers, 14, is holed up in his bedroom—eyes fixed on his computer screen—where he has been logged onto a MySpace chat room and AOL Instant Messenger (IM) for the past three hours. His twin sister Bronte is planted in the living room, having commandeered her dad's iMac—as usual. She, too, is busily IMing, while chatting on her cell phone and chipping away at homework. By all standard space-time calculations, the four members of the family occupy the same three-bedroom home in Van Nuys, Calif., but psychologically each exists in his or her own little universe.

The gadgets and brand names change over time, but the tendency remains the same: away from the few and the near, toward the many and the far. Parents, the magazine concluded, should teach their kids "that there's life beyond the screen." In fact, most parents don't need to be told that, and many have been trying for years. They aren't having much success because our thinking has never gotten beyond the vague notion that "there's life" of some unspecified sort out there that's good for you, kid, trust us, and you'd better go find some now. This is the old eat-your-brussels-sprouts argument that's never worked for any generation, and it's a particularly weak approach to this problem.

Kids aren't stupid, and they're especially good at spotting double standards. Everything they see and hear around them tells them that the screen is where all the fun and action are and where they need to go to thrive and succeed. The occasional news report tut-tutting digital addiction can't undo a thousand others touting the new "must-have" gadget, the social network *everyone's* joining, and so on. Parents can lecture all day, but their moral authority is rooted in their own lives. What can Mom and Dad know about this alleged life beyond the screen if they themselves never go twenty minutes without a BlackBerry glance?

The Nielsen Company reported that in one three-month period, American teenagers sent and received an average of 2,272 text messages each per month, which was more than twice as many as a year earlier. This was viewed as shocking news, seized on as the cause of rampant distraction in school, failing grades, and numerous other ills. What's far more shocking is that we were shocked at all. *Of course* children are texting like crazy. Of course they spend so much of the day huddled with screens, they're barely aware of the third dimension (true headline: "Teen Girl Falls in Open Manhole While Texting") and increasingly unfamiliar with the natural world—nature-deficit disorder, it's now being called. This is how we grown-ups are teaching them to live, implicitly and explicitly, with a conviction they can't fail to miss.

Educator and writer Lowell Monke shared with his students

a troubling study that showed that many young people prefer to interact with machines rather than directly with human beings. The next day, one of the students sent him an e-mail explaining why this might be:

> I do feel deeply disturbed when I can run errand after errand, and complete one task after another with the help of bank clerks, cashiers, postal employees, and hairstylists without ANY eye contact at all! After a wicked morning of that, I am ready to conduct all business online.

"In a society in which adults so commonly treat each other mechanically," Monke writes, "perhaps we shouldn't be surprised that our youth are more attracted to machines." We believe in our screens so much, we've placed them at the center of our lives, so why shouldn't they? If anything, the kids deserve merit badges for doing their best to emulate the values and norms of their community and their elders—to be more like us.

For years, conventional wisdom held it was the young, the so-called digital natives, who were leading the way into the connected future, with grown-ups reluctantly tagging along. This notion was based on hard statistics about relative technology usage by various age groups and an abundance of anecdotal evidence. Younger people are always comfortable with the new technologies of their own era because to them the devices aren't "new" in the way they are to those who can remember a world without them. Kids took digital screens in stride just as their parents took TV screens in stride decades earlier: the gadgets were there in front of them, and they did interesting things—what's the big deal? However, as with television fifty years ago, today's children didn't purchase the first screens they encountered as toddlers. This revolution was started by grown-ups, and if many older people were initially slower to adopt the digital life wholesale, they've played an excellent game of catch-up. By 2009, people over thirty-five were driving the growth of then-cutting-edge digital tools such as Twitter, giving the lie to the youth paradigm.

In the end, this isn't about any one generation. The girl who sent 300,000 texts in a month didn't make the news because she was young or some kind of freak. She made the news because she represented, in slightly exaggerated fashion, how everyone, regardless of age, now lives. When you hear a middle-ager bellyaching that "these kids" don't make a move without their screens and barely know how to conduct a face-to-face conversation, they're really talking about themselves. We've all immersed ourselves in one very particular mode of connectedness, to the point of obsession and pulled away from all other modes. Why? Because to share time and space with others in the fullest sense, you have to disconnect from the global crowd. You have to create one of those gaps where thoughts, feelings, and relationships take root. And for a good maximalist, there's nothing worse than a gap.

IF THERE'S ANYWHERE one would expect maximalism to have no downside whatsoever, it's in the most outward dimension of life, the bustling world of work and business. The free-market society is itself an ingenious form of connectedness, one in which the goal is to sell goods, services, and ideas to as many people as possible and reap the rewards. To thrive in the marketplace, businesses and other organizations are constantly seeking competitive advantages, in technology above all. From the start of the digital age, it's been management gospel that an office can't be too connected. The more wired an organization and its workers are, to each other and the world beyond, the better positioned they are to compete and thrive. In other words, the pursuit of excellence requires the pursuit of connectedness.

Lately, however, it's become clear that it's not so simple. What's true in our individual lives and families is equally true in the workplace: the tool that giveth also taketh away. Once again, it all comes down to what digital busyness does to the mind. These gadgets are adept at performing various different tasks simultaneously and switching quickly among them. As I wrote this sentence on my laptop, for instance, in addition to the word-processing document I was focused on, there were seven other applications open, plus dozens of internal processes working in the background. When I clicked away briefly from this text just now to check my e-mail, the computer deftly made the switch, and just as deftly switched back. One second it was crisply displaying my words as I typed them, the next it was showing me what was in my inbox, and then (since the e-mail I was waiting for hadn't arrived) it returned immediately to the words, with no perceptible loss in performance.

The human mind can also juggle tasks, of course, which explains why you can sit in a café and read a book, taste the coffee you just sipped, and hear the pleasant music playing in the background, all at the same time. However, we can only *really* pay attention to one thing at a time. If the book is gripping, the music will fade into the background of your consciousness, and you'll forget to sip the coffee, discovering a half hour later that it's gone cold. And, unlike computers, when we switch tasks—either by choice or because we're suddenly interrupted—it takes time for our minds to surface and focus on the interruption, and then still more time to return to the original task and refocus on *that*.

Psychologists tell us that when you abandon a mental task to attend to an interruption, your emotional and cognitive engagement with the main task immediately begins to decay, and the longer and more distracting the interruption, the harder it is to reverse this process. By some estimates, recovering focus can take ten to twenty times the length of the interruption. So a one-minute interruption could require fifteen minutes of recovery time. And that's only if you go right back to the original task; jam other tasks in between and the recovery time lengthens further.

Returning to the café, let's say that while you're reading that great book, a friend stops by to say hi. Just as you begin chatting, your phone rings and you ask the friend to hold on a second while you answer it. As you're taking the call, the waitress interrupts and asks if you want a refill. While she's holding the carafe over your mug awaiting an answer, the café's fire alarm goes off. In a matter of minutes, you've gone from three potential objects of interest (book, music, coffee), with one squarely at the center, to seven potential objects (book, music, coffee, friend, phone, waitress, fire alarm), with *none* at the center. Satisfying immersion has given way to unsatisfying confusion. Even when things have calmed down again, the spell is broken and you might as well forget about the book.

What does this have to do with offices and technology? These two café scenarios represent what's happened to the American workplace in the last several decades, as screens have added countless tasks and distractions to every cubicle. The office worker of 1970 had numerous responsibilities and tools to manage, including multiline telephones that had to be answered when they rang because there was no voice mail. Still, it was a relatively disconnected world, and the array of competing tasks was much smaller than it is today. So it was easier to choose one and stay with it, while others waited quietly in the background. Today, thanks to our screens, as we work we're constantly contending with far more tasks than our minds can handle. We find it increasingly hard to concentrate on any one of them for more than a few minutes. It's estimated that unnecessary interruptions and consequent recovery time now eat up an average of 28 percent of the working day. In cubicles everywhere, daily existence now mirrors the book-music-coffee-friend-phone-waitress-fire-alarm onslaught, all the time.

When it's happening live, the seemingly harmless journey from click to click to click, it doesn't feel all that significant. You look away from what you're doing every five minutes to check the inbox—so what? To grasp why it matters, you have to think not so much about what we're doing when we click as what we're not doing. In the first place, we're not working as efficiently as we could, because of the time that's wasted as we lose and then regain focus over and over. Digital work appears to happen at lightning speed, but only because we conflate the speed of our gadgets with the speed of our thoughts. In fact, it's the way screens allow us to shift rapidly *among* tasks that winds up slowing down our execution of the tasks themselves, due to the recovery problem. It's a false efficiency, a grand illusion.

And inefficiency isn't the worst of it. When work is all about darting around screens, we're *not* doing something that's even more valuable than thinking quickly: thinking creatively. Of the mind's many aptitudes, the most remarkable is its power of association, the ability to see new relationships among things. The brain is the most amazing associative device ever created, with its roughly 100 billion neurons connected in as many as a quadrillion different ways—more connections than there are stars in the known universe. Digital devices are, in one sense, a tremendous gift to the associative process because they link us to so many sources of information. The potential they hold out for creative insights and synthesis is breathtaking. The best human creativity, however, happens only when we have the time and mental space to take a new thought and follow it wherever it leads. William James once contrasted "the sustained attention of the genius, sticking to his subject for hours together," with the "commonplace mind" that flits from place to place. Geniuses are rare, but by using screens as we do now, constantly jumping around, we're ensuring that all of us have fewer ingenious moments and bring less associative creativity to whatever kind of work we do.

Yet even though the tools designed to make the workplace more efficient and workers more productive are having the opposite effect, businesses continue ramping up their connectedness, unable to shake their faith in the maximalist approach. One of the sharpest observers of digital life, syndicated cartoonist Jen Sorensen, captured the absurdity of this cycle in an installment of her *Slowpoke* strip entitled "Small Business Meets the Virtual Vortex." In the first panel a businesswoman is shown eagerly taking an order on the telephone. "A dozen by noon?" she says. "You got it!" The caption: "In the beginning, you did your work, and it was good." The second panel reads, "Then you needed a website," and we see our entrepreneur at her screen, proudly launching her business's new online site. In the third panel she's adding a blog. Next she's joining social networking sites and, when that isn't enough, posting short updates about her status all through the day, such as "Don't miss my tweet from 11:27 am today!" In the subsequent panel, she's staring at her screen with a look of confusion. "Wait a minute . . ." she muses, "I forget what I do for a living!!" At the end, two aliens are shown peering down at Earth from a spaceship. "HA HA!" they're saying. "The humans will soon cease all productive activity, and then we can invade!"

For years, businesses didn't see this problem or pretended it wasn't there. It finally got their attention when it started showing up in the bottom line. One study by Basex, a leading research firm that focuses on technology issues in the work-place, found that workers were spending more than a quarter of their day managing distractions. As a result, the firm concluded, businesses were seeing "lowered productivity and throttled innovation." In 2009, Basex estimated that information overload was responsible for economic losses of $900 billion a year.

As this and similarly shocking data emerged, the technology industry, typically viewed as one of the great engines of prosperity, has found itself playing the uncustomary role of economic villain. Having created the tools that got us into this pickle—and, not incidentally, having watched its own ultra-connected workforces struggle with it daily—it realized the onus was on it to do something. A few years ago, concerned executives from some of the largest tech companies, including Microsoft, Google, Xerox, and Intel, got together with academics, consultants, and other interested parties and formed a nonprofit called the Information Overload Research Group. Its mission was to raise awareness and come up with solutions to "the world's greatest challenge to productivity." The *New York Times* reported the news of the group's founding on its front page, under the headline "Lost in E-Mail, Tech Firms Face Self-Made Beast."

So much of our existence is wrapped up in our efforts to earn money and stay afloat that it sometimes seems as though life is just about business. But the challenge we're talking about here isn't ultimately economic or organizational; it's a human challenge that affects everything modern humans do. The most compelling indicator of its seriousness isn't a dollar figure or a productivity statistic but a signal coming loud and clear from the place where all human industry begins: inside our heads. There's a feeling, an impulse that surfaces regularly now in all kinds of situations, personal and professional. It's the desire for a reprieve, a break from the digital crowd.

It's in the weary comments of friends, neighbors, and colleagues about overflowing inboxes and children who can't be pried from their screens. It's in the immense popularity of yoga and other meditation practices, which now serve as useful, albeit temporary, respites from digital busyness. It's in the Slow Life movement—slow food, slow parenting, slow travel—with its valuable message that everything has simply gotten too fast.

It's in the quiet car on the train and the sign at the cash register that says, "We will be happy to serve you when you have completed your cell phone conversation."

It's in mobile phone throwing, an international "sport" invented by some puckish Finns as "a symbolical mental liberation from the restraining yoke of being constantly within reach." The annual world championships draw whimsical media attention, but the stories tend to leave out a telling fact: Finland ranks among the most connected countries on Earth. It seems that the more connected we are, the heavier the yoke.

Not many of us can afford to throw away our tools, so the next best thing we can do is run away from them. Thus the burgeoning interest in "unplugged vacations" and travel to extremely remote places. "You won't find a television, telephone or WiFi outlet in any of the 22 cottages at Petit St. Vincent, an idyllic resort on its own private island in the Grenadines," reports a business-travel magazine. "What you will discover are stretches of empty, pristine white beaches sprinkled with hammocks and shady trees and a plethora of quiet nooks to cuddle with a good book." As it happens, people can still use their smart phones on this island—in a wireless world nothing is as private as it seems—but the owner calls that "a terrible mistake." On a different island, a resort has a creative solution to this problem, an "Isolation Vacation" package that offers "a seven-night getaway" for $999, with the requirement that on checking in you turn over your mobile devices to the staff, to be locked up for the week.

With connectedness approaching ubiquity, physical remoteness no longer ensures isolation. This is an important shift we haven't quite wrapped our minds around. I see it when friends who live in bustling metropolitan areas tell us they envy our "disconnected" life in a far-flung place. One of them, a New Yorker who visits the Cape every summer, complained to me about the endless hours her children spend messaging and gaming on their screens back in the city. "You're lucky," she said, "you don't have this problem on Cape Cod." Oh no? It doesn't matter if home is a noisy urban walk-up or a quaint cottage on a secluded bluff. If you have a screen and can pick up a signal, your mind is in the same placeless place. Still, the romance of the away-from-it-all life persists, as if we're trying to wish it back into being.

Wishing isn't enough. Imagine you woke up one morning and found that one of your most treasured possessions—say, a painting that had always hung on a certain wall and made you happy every time you looked at it—had been stolen. You'd want it back, of course. Would you just dream about recovering the painting, in the hope that dreaming alone would magically restore it to the wall? Or would you take action? We gab endlessly about the relative merits of our various digital devices, which ones are faster and easier to use, and barely utter a peep about what they've spirited away and how to regain it. It's hard to talk about this missing possession, because it's intangible and amorphous. It's an absence that we miss, of demands and distractions, rather than a presence. How do you get back something you can't even describe?

We're all in the same boat, and what we need is what I found by accident when I fell out of mine. The problem is, we also want and need our devices. A few days after my *What Larks!* mishap, I bought a new mobile phone. Of course I did. I wasn't interested in being cut off from the world, and, practically speaking, I couldn't be. Besides, my phone is also a source of happiness.

Spectacular benefits and enormous costs, in the very same tools. If we could just grow the former and shrink the latter, the potential of life in this connected world be boundless. Screens would be instruments of freedom, growth, and the best kind of togetherness, as they should be. The question is how.

Chapter Four

SOLUTIONS THAT AREN'T

The Trouble with Not Really Meaning It

Shortly after the technology executives founded the non-profit Information Overload Research Group, with the explicit goal of fighting excessive digital connectedness, Microsoft launched a new ad campaign. It featured Bill Gates, Microsoft's chairman, in a series of humorous skits with the comedian Jerry Seinfeld, touting the company's technologies.

"Bill, you've connected over a billion people," Seinfeld said in one of the sketches. "I can't help wondering what's next. A frog with an e-mail? A goldfish with a Web site? Amoeba with a blog?"

Gates suggested he wasn't too far off. The screen then went blank except for a two-word slogan: PERPETUALLY CONNECTED. Not moderately or rather connected, not connected within reason or as needed, but perpetually connected, as in ceaselessly, without pause. At the same time that the tech industry was garnering major media attention and plaudits for finally recognizing that spending all of one's time connecting via screens is a terrible idea, Microsoft turned around and offered that same idea as its mission and, by implication, ours.

This sort of thing happens all the time with industries whose products can be addictive or otherwise hazardous. The alcoholic beverage industry stands foursquare against alcoholism and says so in public service campaigns, while it simultaneously spends billions of dollars encouraging us to drink. With booze, at least there's a genuine distinction between pushing alcohol, which the industry does, and pushing alchol*ism*, which it doesn't. In promoting a lifestyle of never-ending connectedness, the technology business (and Microsoft is hardly alone in this) is encouraging the unhealthy extreme, the digital equivalent of alcoholism. Perpetually connected equals perpetually zoned out, and nobody knows it better than those companies. In studies of workplace overload, the most shocking statistics and anecdotes—employees so distracted, they can barely think—come from the technology sector.

The same digital doublethink prevails in the media, a business that's supposed to tell it like it is and clear up confusion on important public issues. For years now, news outlets have been dutifully covering what's been aptly called the Too-Much-Information Age. Though rarely given prominent play, these pieces are a minor staple in even the most tech-positive outlets. Wired.com, for instance, warned in a headline that "Digital Overload Is Frying Our Brains."

At the same time, the news media push nonstop connectedness as eagerly as any Silicon Valley titan, and for the same reason: business demands it. They need an audience for their product, and the bigger the audience, the better. If 2 million people are visiting your Web site once a day for ten minutes each, that's very nice. If those same 2 million are glued to it around the clock, clicking constantly on your latest content and ads, breaking away only for meals and bathroom runs, that's excellent! You're in the money! Hence the same outlets reporting on the perils of overload also promote it, sometimes simultaneously.

"Warning," began a piece by the *Wall Street Journal*'s "Information Age" columnist L. Gordon Crovitz. "On average, knowledge workers change activities every three minutes, usually because they're distracted by email or a phone call. It then takes almost half an hour to get back to the task once attention is lost. . . . Consider the rest of this article an 800-word test of your ability to maintain attention." It was a smart piece, articulating well the case for a new approach while praising the tech leaders for their nonprofit effort. "It's encouraging that the most information-intense companies are trying to overcome their own overload." As I read this in the paper's digital edition and tried to meet the 800-word challenge, my eye was drawn to a colorful box on the right side of the screen, where a "house ad" (from the paper itself) was flashing the message "STAY CONNECTED 24/7 VIA EMAIL NEWSLETTERS & ALERTS . . . FREE Registration. Sign up Today." That is, *Click here for more overload.* Columnists have no control over ad copy, but for the reader, the dissonance is hard to miss, an echo of one's own internal conflict.

On the morning public radio show *The Takeaway*, the host John Hockenberry asked listeners to share how they escape the distractions of their gadgets. "Tell us your stories of focus," he said, offering a Web address for posting of same. So if, like me, you happened to be focused quite nicely on this broadcast—nothing dovetails better with making breakfast than a good radio show—you were supposed to find the nearest screen and thereby dilute that focus, in order to share with the world . . . how you focus! For a moment I actually considered it. There was a screen steps away, and I was well aware that while tapping out my thoughts on focus I could dart off and peek at my inbox. You know, just to check. "I think we're all in danger of being on a short cell phone leash," said one listener who called in a comment, presumably using his cell phone.

So it goes, the dog endlessly chasing its tail. And it's easy to see why. What's a technology giant supposed to do, run ads urging the public to ease off on its technology habits? Of course not. No news organization in its right mind is going to say, "Come to our Web site—but not too often, okay?" And it's a very good thing that a radio show is having a discussion about focus. We *should* talk more about this and share ideas, and what better way to share them than digitally? The screen is by far the best place to get a message out these days, and there's nothing wrong with that. If your message is that people are going overboard with screens, then that's where you need to be, because it's where the most afflicted are.

We're all speaking out of both sides of our mouths on this issue. When the word "CrackBerry" was in vogue a few years ago, those who uttered it most often and most bitterly were the addicts themselves. The question is why one of the two sides—that we're all overdoing it with screens—is not having any appreciable effect.

It's not as if nobody's trying. As awareness of the quandary has grown, so has the search for solutions. Here again, the business world has been leading the way because, as one IBM researcher put it, "There's a competitive advantage of figuring out how to address this problem." So far the ideas have come in two basic varieties. The first is old-fashioned time management, the notion that one can impose order on the digital chaos by setting aside certain times of day, or certain days of the week, for particular tasks. For example, there are people who limit their e-mail checks to fixed hours of the day, say at 9 A.M., 1 P.M., and 5 P.M. Some businesses have applied the time management approach broadly through companywide screen hiatuses, such as no-e-mail Fridays. The aim is not just to discourage excessive screen time but also to encourage face-to-face interaction, which is often more efficient and productive than long, multi-recipient e-mail chains. If everyone is disconnected at the same time, they're more likely to come out of their cubicles and talk.

Despite numerous experiments along these lines, time management solutions haven't taken off. In many cases, workers simply cheat to get around the restrictions. The reason it's hard to make such methods work is that they're essentially diets, but rather than counting calories they count screen hours. And like all diets, they seem far more feasible in the abstract than in practice. In a world where everyone is gorging on connectedness, it takes serious willpower to say, "None for me today, thanks." Besides, screens are at the heart of most people's professional *and* personal lives, and taking them away, even for half a day, means falling behind on all fronts.

The second approach looks to technology itself as the answer. Here the solutions range from simple software switches, which allow anyone to shut off her inbox or signal others that she's currently unavailable, to elaborate filters and "digital assistants" programmed to assess the relative importance of incoming messages and keep trivia at bay. These, too, have been around for years, in some cases available for free online. Yet how many of us are using them?

The technological approach has a couple of weaknesses. One is that it typically focuses on the symptoms—too many messages and other tasks on the screen—without touching their sources. It's nice to see only the messages that your e-mail robot has rated important, but that doesn't mean the lesser ones cease to exist or won't have to be dealt with later. Nor will it stop the world from sending you more. And though filters reduce the apparent supply of tasks, they do nothing about the demand you generate yourself. Let's say you install an e-mail filter that successfully hides your low-priority incoming traffic, gaining you an average of thirty minutes a day. What's to stop you from spending that extra time creating new outgoing messages of your own or just going off to idly check headlines or your stock portfolio or that baseball blog you're addicted to? Connectedness begins at home, and, let's face it, we're our own worst enemies.

Another weakness of the technological route is that it's based on the often erroneous assumption that laborsaving devices actually save labor. If the digital era has taught us anything, it's that a new technology frequently creates more work than it saves. Once you've installed a digital assistant to monitor your e-mail, who's going to monitor the assistant—adjust the settings, clear out the rejects file, update the software regularly, and perform all the other time-intensive housekeeping chores that screen life requires? Assuming you don't have a staff of assistants at your beck and call, I think you know who those duties will fall to.

Some of the technocures for overload seem designed to make it worse. There's software that can sense a worker's keyboard and mouse activity and use this to gauge when he or she can be interrupted. The assumption is that if you're not tapping or clicking, you aren't doing anything important. Just sitting there thinking doesn't count as valuable work, though it's in that state of aimless reverie that the best ideas tend to come, the prized "eureka" moments. Another widely touted idea is to increase the rate at which we process our digital information, cramming more content into each minute. For instance, there's a gadget that displays e-mail messages one word at a time "for accelerated reading speeds that can reach up to 950 words a minute." Perhaps this is possible, but does anyone seriously believe it aids thinking or reflection?

As if to acknowledge that more technology is not the answer, some have suggested we instead off-load our digital burdens onto other human beings. A popular self-help book recommended the outsourcing of e-mail and other drudge work to paid assistants in the developing world, as the author himself did by hiring a few such people in India. "It's the fourth morning of my new, farmed-out life," Timothy Ferriss writes, "and when I flip on my computer, my e-mail inbox is already filled with updates from my overseas aides." Again, note the bias toward busyness, the crowded inbox as the measure of success. It also reads like some back-to-the-future version of the British Raj. *Your downloads are ready, Sahib.* So much for technology as liberation.

All of these efforts are aiming for the same worthy goal—a saner work life—and all have the same basic flaw. They're seeking outward answers to an inward problem. Our busyness doesn't just take place in our minds, it's our minds that orchestrate it and allow it to happen. When anyone mentions the mind today, most of us immediately think of the brain, though they're not the same thing. When I talk to friends about the challenges of connected living, nine times out ten they bring up neuroscience, a field that's been exploding in recent years, thanks to imaging technologies that allow researchers to observe the brain in action. Today there's no more impressive lead-in than "According to a new neuroscience study . . ." Perhaps, the hopeful thinking goes, the answer will be found there.

There's a great deal of justified excitement about the latest brain research, much of it built on decades of previous work. On the specific question of how the latest digital technologies work on the brain, however, there's no such body of knowledge, because the devices are so new. Hence the research is still very preliminary, the findings tentative.

Part of what drives us back to the screen may be evolutionary programming. The human brain is wired to detect and respond to new stimuli. When we become aware of some novel event or object in our surroundings, the brain's "reward system" is activated, which involves the use of neurotransmitter molecules in the form of dopamine. Some researchers

theorize that this is all a bequest from our prehistoric ancestors, whose survival in a dangerous world depended on their
ability to perceive threats (such as predators) and opportunities (a potential meal) in their immediate surroundings and
respond quickly. Today the stimuli we receive from our environment are different—instead of wild animals lurking in the
trees, we're on the alert for ringtones and new messages—but
the biochemical effect is hypothetically the same. When your
mobile lights up with a new call, you get, in the words of one
scientist, a "dopamine squirt."

Of course, there's a crucial difference between 100,000
years ago and today. In the primitive world, where life moved
more slowly, it would have made more sense to have your attention managed that way. Our survival doesn't require that
we pay attention to all the new information that now comes
our way day and night via screens. A viral video can't eat you
for dinner the way a lion can, and if you ignore the e-mail that
arrived three seconds ago, the odds are quite good you'll live
another day. Yet, as we all know, the urge is hard to resist. In
effect, the theory goes, by feeding us a constant stream of distractions and novelties at shorter and shorter intervals, our devices take advantage of certain ancient brain structures. This
might explain that nagging sense that the screen urge is not
entirely rational but preconscious and automatic.

It's possible that our brains will eventually adapt to a digital
world and learn to better manage all these pulls on our attention. The organ's plasticity, or ability to change by rewiring itself, is well known. However, neuroplasticity is not the
panacea it is sometimes made out to be. There are fundamental limits on our attentional capacity, based on the amount of
brain space we have for what's called working memory. For
that to grow would require a *structural* change far more significant than the rewiring of neural pathways. So, despite the

touted benefits of the various "brain training" gadgets now being marketed as solutions to attention problems, it isn't that easy.

Besides, it's not ultimately the brain that's overloaded; it's the thoughts and emotions that somehow arise within the gray matter—consciousness, the mind.

Brain and mind are intricately related, but in ways we've barely begun to understand. "We are still clueless about how the brain represents the content of our thoughts and feelings," writes psychologist Steven Pinker. It's the mind that defines our lives, and we know quite a bit about how the mind works. Antonio Damasio, the pioneering neuroscientist who coined the phrase "movie-in-the-brain" to describe human consciousness, has observed that there is currently a "large disparity" between our knowledge of how the brain works, which is incomplete, and "the good understanding of mind we have achieved through centuries of introspection and the efforts of cognitive science."

It's not just the hardware that matters but the software, the ideas. People change their behavior when they embrace a new way of thinking about it. Whether it's acquired from reading, therapy, a twelve-step program, or some other source, the philosophical approach isn't merely a control mechanism, a way of tamping down persistent urges. At its best, it's a profoundly creative force, a way of rethinking and reshaping some important aspect of life that's giving us trouble.

What's giving us trouble right now is our screens. We've been saying we need to change our habits, but, as our behavior makes clear, *we don't really mean it*. If we meant it, we'd be living differently. To mean it, you have to believe it, and to believe it, you need a set of convincing ideas. Those ideas might not alter the physical structure of our brains, but that's not necessary. As long as they change our minds, that will be enough, because our behavior will naturally follow. People aren't going to change deeply ingrained habits just because company policy says it's bad for the bottom line or because a software program has decided to block the inbox. But if there's something valuable to be gained, more valuable than what's gained by staying glued to the screen, they just might. If the many failed solutions to the problem of overload were grounded in compelling principles and goals, some of them could actually work.

The twentieth-century thinker Michel Foucault had a nice phrase for philosophical tools that help us improve and transform our lives: technologies of the self. That's what we need now, a new technology of the self for a digital world.

Here's a place to begin:

> Turn off your computer. You're actually going to have to turn off your phone and discover all that is human around us. Nothing beats holding the hand of your grandchild as he walks his first steps.

These words didn't come from a Luddite malcontent trying to hold back progress. The speaker was Eric Schmidt, the chairman and CEO of Google, in a commencement address delivered at the University of Pennsylvania in the spring of 2009. In these hyperconnected times, it would have been an arresting statement no matter who had made it. But coming from the head of the organization that has, more than any other, defined the new connectedness, it was a stunner. Google isn't just a search engine, it's a vast media and advertising company whose profits are tied directly to the screen habits of people around the world. In urging the young graduates in his audience, and by extension everyone else, to turn off their computers, Schmidt was promoting behavior that would be demonstrably bad for his own bottom line.

A cynic would say he was just playing the lofty orator, knowing full well that our digital fixation runs so deep, his idealistic counsel would have no real effect. That, like other tech figures who have ostensibly backed the fight against overload, he was just talking the talk. But why bother to say it at all—he could have spoken about any topic—and why so directly and earnestly? "Turn off your computer" and "discover all that is human around us" are very different kinds of statements from "Let's develop industry solutions to information overload." You don't often hear language like this from someone in Schmidt's position. But the world being what it is, the quote made the media rounds for a day (Tech mogul says we should disconnect, ironic!) and poof, it was gone.

What's interesting about it isn't so much the specific advice Schmidt was imparting—"Just disconnect" is hardly an original thought—but the ideas underpinning it. He was suggesting that for all the good work our screens do for us, there are some experiences they can't deliver, and those happen to be the most important ones. He didn't use the word "depth," but that's what he was saying is lost when we live on and through the screen. His simple exhortation to turn off the screen assumed that we each have the capacity to recognize this and the power to change how we live with these devices. And that it's up to us *as individuals* to use that power. At a time when the crowd is held up as the source of all authority and meaning—we Google ourselves to see if we matter—those are radical thoughts.

In practical terms, he was saying everyone needs to create a gap between themselves and the screen—the gap that opens up when you turn it off. When you do that, something miraculous happens. You regain the best part of yourself and the best part of life, the human part. Previous efforts to solve this conundrum, whether by diets or devices, have also been premised on the need for a gap. But they were missing the inner incentive, the reason to believe that Schmidt pointed toward.

We all regularly have experiences that point in the same direction, though we generally don't see them that way. When my phone call with my mother ended, I opened up a gap that took me to the same place he was talking about. In my case, the digital experience itself was part of my journey back to "all that is human." But it wouldn't have happened if I hadn't ended the screen interaction. Ideally, we'd be able to find a set of ideas that would help us do this on a regular basis, so we'd get healthy doses of both kinds of experience, and each dimension would enrich the other. If we traveled back and forth more often, the digital zone would inevitably become more human. And isn't that really the goal?

The kinds of ideas I'm talking about, new philosophical approaches to the screen, are available to us right now in a somewhat unlikely place: the past. The "centuries of introspection" that have produced so many insights into the human mind and how best to use it also have a great deal to teach about the mind's relationship with technology. Though it often seems as though we're living in a completely new age, unlike anything that came before, this technological transformation has many precedents. History is replete with moments when some astonishing new invention came along that suddenly made it easier for people to connect across space and time. And those earlier shifts were as exhilarating and confusing to those who lived through them as today's are to us.

New modes of connecting always create new ways for individuals to create and prosper, and for the collective advancement of humanity. At the same time, there's a sense of life, especially the inner life, being thrown out of balance. It happened in the sixteenth century, after the arrival of print technology, and again in the middle of the nineteenth century, when the railroad and the telegraph appeared. There are many other examples. The mind has been on a long journey, and along the way there have always been a few individuals who have arrived at valuable insights about how best to manage that journey.

The seven philosophers you will meet in the next part of the book lived through eras that resembled our own in certain essential ways. Even though most of them died long before anything resembling modern screens existed, they all understood the essential human urge to connect and were unusually thoughtful about the "screen equivalents" of their respective epochs.

Their lives and circumstances vary widely. One spent much of his life as a struggling entrepreneur. Another was, for a time, one of the most powerful men on Earth. They expressed their ideas in varied ways. Two are routinely ranked among the greatest writers of all time, while another left behind no written record of his thoughts. What they share is a profound interest in the questions raised by human connectedness. What is connecting all about, anyway? What can these tools do for us? What are their strengths and weaknesses? How can we use them to build better, more satisfying lives?

The past is not the conventional place to look for guidance on the new connectedness. In this forward-drive world, history can feel like the computer you owned fifteen years ago, outdated and pointless. What can seven dead white guys possibly teach us about life in a rapidly changing global society? More than you might imagine. Technology and philosophy are both tools for living, and the best tools endure and remain useful over long periods of time. Though we barely realize it, every day we use connective tools that were invented thousands of years ago. Similarly, great ideas have no expiration date.

What's most striking about these figures is how modern they often seem. As users of the "screens" of their day, they felt a pull much like the one we feel. At the same time, they and their contemporaries yearned for all that we yearn for: time, space, tranquillity, and, above all, depth. It's as if they saw the future coming and, in a way, lived it. The world has changed hugely over the centuries, but the basic ingredients of human happiness haven't.

PART II

BEYOND THE CROWD

Teachings of the Seven Philosophers of Screens

Chapter Five

WALKING TO HEAVEN

Plato Discovers Distance

"It's more refreshing to walk along country roads than city streets."

One of Plato's greatest dialogues begins in Athens on a beautiful summer day. It's the late fifth century B.C., a period sometimes referred to as Greece's golden age because there were so many gifted artists, poets, playwrights, philosophers, and statesmen living and working there at the same time. One of the most famous of them, Plato's teacher Socrates, spots a young man he knows walking down the street and calls out to him.

"Phaedrus, my friend! Where have you been? And where are you going?"

This hearty greeting captures the essence of Socrates, a man who cherished his friends and was intensely curious about their lives. He was an avid connector in the face-to-face sense, a trait that comes out over and over in the philosophical conversations he had with fellow Athenians, which form the backbone of Plato's writings.

This dialogue, known simply as *Phaedrus*, explores human connectedness in a time of dramatic technological change. A revolutionary new form of communication, written language, had arrived in Greece, which had long been an oral society. It was beginning to catch on, and thoughtful people were worried about its effect on various aspects of life, particularly the life of the mind. In other words, though this story takes place roughly 2,400 years ago, it's about an era somewhat analogous to our own. Writing on the cusp between two technological eras, Plato examined questions that are in the air once again today.

Phaedrus tells Socrates that he's just spent the entire morning with the well-known orator Lysias, listening to his latest speech. To the modern reader, this might seem a curious way for a young man to be spending his time, but in a society largely organized around the spoken word, it was perfectly natural. Just as social networks and viral video clips are all the rage today, in rhetoric-obsessed Greece there was nothing cooler than sitting at the feet of a brilliant speaker, soaking up every word.

The speech was about a topic that's always of urgent interest: sex. Specifically, it was about the question of whether it's better to sleep with someone who's in love with you or someone who isn't. Lysias argued for the latter position, pointing out that when you have sex out of pure lust there are far fewer emotional complications.

Phaedrus thought the speech ingenious, and he's been walking around turning it over in his mind, trying to commit it to memory. In pursuit of this goal, he's headed outside the city walls, following the advice of a prominent doctor named Acumenus that "it's more refreshing to walk along country roads than city streets." He invites Socrates to join him and hear more about the speech, and the older man readily agrees. They set off, eventually leaving the footpath to walk barefoot through a stream. They follow it until they reach a beautiful spot beside the stream where they can sit under a plane tree and talk.

Socrates marvels at what a lovely, tranquil place it is, prompting Phaedrus to observe that the philosopher seems like a total stranger to these natural surroundings: "As far as I can tell, you never even set foot beyond the city walls."

Socrates concedes that it's true. "Forgive me, my friend. I am devoted to learning; landscapes and trees have nothing to teach me—only people in the city can do that." He's journeyed here, he says, strictly because Phaedrus enticed him with an invitation to do what he loves to do best back in Athens, talk over a philosophical question like the one addressed by the speech. With that, he lies down on the grass and asks Phaedrus to recite Lysias's arguments for no-strings-attached sex.

When was the last time you went off with a friend and truly left the rest of the world behind? Socrates and Phaedrus are enjoying a type of human connection—in person, dedicated, utterly private—that's quite rare today. Even when you're physically with another person, it's hard to give them your un-divided attention for a sustained period, or to receive the same from them. If there's a digital device nearby, chances are that one or both of you will be distracted or interrupted.

What's interesting is that this secluded chat was a rare ex-perience for Socrates. He admits he hates to leave the crowded city, where his work as a philosopher revolves around his con-versations with students and other intellectuals, usually in larger groups. In fact, this is the only one of Plato's many dia-logues in which Socrates leaves Athens for a private tête-à-tête.

The philosopher had a deep craving for the oral connect-edness that was dominant in his time. You might say he was an ancient maximalist and Athens was the "screen" that en-abled his habit. Now, like a modern road warrior with a mobile broadband device, he's ventured into the hinterland in the hope that he'll be able to find a good connection there, too. And he's waiting for Phaedrus to provide it, with a rendition of that libidinous lecture. Though life in ancient Greece was obviously different in many ways from life in the twenty-first century, the basic human desire to connect was the same. Socrates was seeking what everyone with a digital screen is after: contact, friendship, stimulation, ideas, professional and personal growth.

This outward urge goes back much further than the fifth century B.C. Countless thousands of years ago, our prehistoric ancestors knew nothing about the world beyond their immediate surroundings and had no connective tools with which to transcend their isolation. In fact, there was a time in the very distant past when they couldn't even converse with their closest companions, because they didn't know how.

Somewhere along the way, nobody knows exactly when, an amazing thing happened—or rather, two amazing things. Prehistoric humans came up with two of the most powerful connective tools ever devised, as E. H. Gombrich recounts in his book *A Little History of the World*:

> They invented *talking*. I mean having real conversations with each other, using words. Of course animals also make noises—they can cry out when they feel pain and make warning calls when danger threatens, but they don't have names for things as human beings do. And prehistoric people were the first creatures to do so.
>
> They invented something else that was wonderful too: pictures. Many of these can still be seen today, scratched and painted on the walls of caves. No painter alive now could do better.

I came across this passage while reading the book to my son at bedtime one recent winter. Gombrich wrote the *Little History* for children, but I learned more from it than I have from most adult history books, because he treats technology and other facets of the past as the human stories they really are, free of specialist jargon and needless complexity. He calls those prehistoric people "the greatest inventors of all time," and he's right. They wanted and needed to reach out beyond themselves, and they found a couple of brilliant ways to do it: words and images.

History retraces this story over and over. People are constantly trying to close the distances between them by inventing new connective tools and working over time to improve them. Humans are the only animals that devise multiple uses for a single tool, and we're especially good at finding new applications for our connective tools. If the "technology" of conversation was originally created to serve the practical needs of people struggling to survive in a harsh environment, by the fifth century B.C. it had evolved into something richer and more interesting: a path to truth and enlightenment.

Socrates used conversation to practice philosophy as nobody had ever practiced it before. Whereas previous philosophers had set themselves up literally as wise men with special access to the truth, he made no such claim. He was "a totally new kind of Greek philosopher," writes modern-day scholar John M. Cooper. "He denied that he had discovered some new wisdom, indeed that he possessed any wisdom at all." Rather, he believed the way to attain wisdom was through searching discussions with others like the ones he presided over in Athens, using the question-and-answer technique known today as the Socratic method. For Socrates, oral communication was the key to a good life.

But there was a downside to the connectedness of oral society. Talking enabled the emergence of early civilizations like Greece and the cities that were their nerve centers, none of which would have been built if people couldn't communicate their thoughts. These ancient metropolises offered many benefits to those who lived in them, including the intellectual stimulation that Socrates treasured. At the same time, they imposed new burdens. They were busy places, not anything near as busy as today's cities but, by the standards of their time, busy indeed. To live in Athens was to be surrounded day and night by a few hundred thousand other people, with all their attendant activity, noises, smells, and other claims on one's attention. It was a permanent crowd, and life in a crowd is an inherently demanding experience.

Plato makes it clear that life in Athens could be taxing to the mind when he quotes Phaedrus explaining why he's decided to take a stroll outside the city walls. Like the modern person who takes up yoga or meditation on a doctor's advice, he's following the physician Acumenus's prescription for clearing the head. He's getting a little exercise, and in a very particular way. In order to think deeply about the speech, he's putting some distance between himself and the crowd.

Distance. The very thing human beings had been running away from since prehistoric times, the space separating the self from others. The point of oral communication and all the good things that flowed from it had been to shrink the distances between people. Now, in the place where this kind of connectedness had reached its highest and most intense expression, thoughtful people were realizing that, for personal well-being and happiness, it was necessary to restore some of that distance to everyday life.

This dialogue isn't about distance per se. But Plato was a careful, economical writer, and it's unlikely he would have made so much of the walk in the country unless he was trying to make a point. Phaedrus was a member of Socrates' intellectual circle and, like Plato, deeply interested in rhetoric and philosophy. Thus, when he was walking along in the city trying to memorize the speech, he wasn't just idly musing, he was doing work that mattered to him. And to do it well, he realized he needed some space.

For a twenty-first-century equivalent, think of the cubicle dweller who's spent the entire morning immersed in the digital crowd, shuttling among e-mails, Web pages, text messages, and other electronic activity. She wants to step away and focus on just one thing, perhaps an important project requiring sustained thought and creativity. Though not an aspiring philosopher, this worker is in much the same position as Phaedrus. She's striving to absorb new information, to learn from and make sense of it. But with all that stuff knocking around inside her head, it's awfully hard. How to refresh the overloaded mind?

In Athens, Plato suggested, one answer was creating a physical distance; getting away from the crowd by spending a few hours outside the walls. Curiously, though, Socrates doesn't see the point. He was about sixty at this time, and years of experience had convinced him that conversation was the only reliable path to wisdom and happiness—and the more people who were available to converse with, the better. By this logic, a philosopher (a word that means "lover of wisdom") should never want to put any distance between himself and the crowd. It's the same basic principle that drives digital life today: the more you connect to others through screens, the better off you are.

Who was right, one of the most celebrated thinkers of all time or a young man remembered chiefly as a bit player in that thinker's work? Do we need distance, or don't we? The answer emerges in the balance of the dialogue.

Back at the stream, Phaedrus launches into the speech with the help of a surprising tool. Earlier, just before they stepped into the stream, Socrates said he wouldn't be satisfied with a mere summary of Lysias's argument. He wanted to hear it word for word as originally delivered. Phaedrus protested that he couldn't possibly do that since he didn't have it memorized. Socrates then observed that Phaedrus seemed to be hiding something under his cloak, and he strongly suspected it was a written copy of the speech. At which point Phaedrus sheepishly pulled out exactly that, a hard copy of the oral presentation.

Some translations call it a "book," others a "scroll." Whatever you call it (I'm going with scroll), the point is that, as he headed out for his meditative walk, the younger man had taken with him a tool employing the very latest communications technology, written language based on an alphabet. In fact, writing wasn't completely new. The Egyptians and other early civilizations had pre-alphabetic writing systems. And the Greek alphabet had been around for several hundred years by that time, but it had been very slow to catch on. It was only in the lifetimes of Socrates and Plato that it really took hold. In contemporary terms, Phaedrus's scroll was roughly what a cell phone was around the year 1985, a technology still in the early stages of adoption and not yet fully understood.

The reason he had brought the scroll along is obvious: it was useful. It would allow him to continue thinking about Lysias's speech and work on memorizing it even as he wandered into the country. With the hard copy in hand, he could engage with the speaker's ideas far from the place where the speech had been first delivered, long after it was over. He could leave the crowded city and still perform the task he wanted to perform. If he's a little embarrassed by the scroll, as he seems to be, perhaps it's because he's in the company of the most revered oral communicator of all time, a man who never read from a written text and, it will soon emerge, didn't think much of the medium.

When Phaedrus is done delivering the speech, Socrates lavishly applauds the performance, playfully pronouncing

himself "in ecstasy." They then have a discussion of its arguments, and along the way Socrates spins one of the most
famous metaphors in all of philosophy. Since Lysias's essential point was that love drives people mad, Socrates examines
exactly what madness is and why the mind sometimes goes
over the edge.

He likens the soul to a flying chariot pulled by a pair of
winged horses. One of the horses stands for the good, virtu-
ous side of us and the other for the bad, corrupt side. The goal
of the charioteer is to drive the horses skillfully so the chariot soars up toward "the place beyond heaven" where "pure
knowledge"—enlightenment and happiness—resides. But the
horses are hard to manage, especially the evil one, and sometimes they pull in different directions. When this happens, the
chariot loses its way and crashes to Earth.

The image still resonates because it captures something essential about the challenge of being human. Socrates aimed to
be a practical philosopher, and what he's describing is really
the journey of the inner self every day. We're all driving our
own chariots through the chaos, struggling to reconcile the
forces pulling at us from every direction. You know the feeling. You rush around chasing the things the world holds up as
the keys to happiness: money, success, status, what passes for
entertainment. Yet they don't do the trick, not in a lasting way.
On some level, you know you could be using your time and
talents to pursue a steadier, more authentic kind of existence,
but you're not sure how. As Socrates puts it, chariot driving "is
inevitably a painfully difficult business."

Foolish people get caught up in the chariot race itself, he
says, "trampling and striking one another as each tries to get
ahead of the others." Others manage to stay calm and keep their
chariots on course, adroitly avoiding the pileups. And while
these lucky souls don't quite attain "pure knowledge"—which

is reserved for the gods—they do soar to impressive heights and find genuine contentment.

Skillful life management yields wisdom and happiness. It's a terrific ideal, but the busier our days become and the more others control the reins, the harder it is to imagine achieving it. Lately, with the relentless demands of digital devices, the challenge often seems insurmountable. If you're a faithful connector who spends all day interacting with screens, you probably know, as I do, what it's like to have your chariot stuck in the bad place. "The result is terribly noisy, very sweaty, and disorderly," Socrates says, and those who live this way wind up "unsatisfied."

What can we do about it? This isn't ancient Greece, and Socrates and Phaedrus never had to manage jammed inboxes. But the beauty of Plato, and the reason he's still widely read today, is that he addresses life's fundamental questions in ways that transcend time and place. The chariot metaphor is a helpful reminder of the link between the outward self—how we spend our time interacting with the world, managing our work lives and relationships—and the inward one. In ancient Athens, there was a highly effective way to quiet one's busy outward life and get the chariot back under control: a simple walk in the country.

True, the star of this story, Socrates, initially pooh-poohs the idea of putting any distance between himself and his beloved city. However, Socrates isn't the only philosopher involved here. Plato wrote this and the other dialogues of Socrates after the latter's death. They're based on real historical conversations, but since time had passed and Plato was becoming a philosopher himself, it's widely assumed he took liberties and often arranged the material to make his own points. Though he never states his personal views directly, now and then he seems to implicitly criticize what Socrates is saying.

Phaedrus is sprinkled with clues that Plato disagreed with his teacher about distance. First there's the question of the walk in the country. Though Socrates leaves Athens reluctantly, once he and Phaedrus have settled in by the stream, they have a conversation that, even by Socratic standards, is extraordinary. After the "ecstasy" of Phaedrus's performance, Socrates delivers a few stunning speeches of his own, becoming so absorbed in the task that he's in a kind of rapture. He's in the zone, you might say, and he attributes this pleasant state to their rural hideout. "There's something really divine about this place," he says. He's using the word "divine" literally, to suggest that the gods are inspiring him. But notice he links the divinity to this *place*, the isolated location to which Plato has devoted particular attention. The message is unmistakable: the distance Socrates had dismissed as a pointless bother has played an important role in helping his mind take flight.

Second, the tool Phaedrus brought along under his cloak allows them to make the most of that distance. With the hard copy in hand, they can be away from town with all its distractions and burdens, yet retain full access to one of its chief draws: great, stimulating rhetoric. The gadget is the linchpin of their conversation, but once again Socrates doesn't see the point.

Toward the end of the dialogue he brings up the new technology and the question of whether written language serves any useful purpose. He tells a story about an Egyptian god named Theuth who had invented many "arts," including arithmetic, geometry, and astronomy. But his greatest discovery was written language. Theuth showed this creation to the king of Egypt, promising it would "make the Egyptians wiser" and "improve their memory."

The king was not impressed. To the contrary, he told Theuth, writing would make his people forget more easily.

Once something was recorded in this external way, using letters, they wouldn't feel the need to "remember it from the inside, completely on their own," i.e., in their minds. Worse, they would use writing to appear knowledgeable when they were merely parroting what they'd read. "[T]hey will be tiresome," the king says, "having the reputation of knowledge without the reality."

Socrates shares the king's dim view of this tool, and he expands on it. Writing is a dangerous invention, he tells Phaedrus, because it doesn't allow ideas to flow freely and change in real time, the way they do in the mind during oral exchange. Whereas conversation is all about back-and-forth, written language is a one-way street: Once a thought is written down, it's frozen and you can't challenge it or change its position. It's a record of ideas that already exist, rather than a way of creating new ones. He likens written texts to paintings, which "stand there as if they are alive, but if anyone asks them anything, they remain most solemnly silent." A piece of writing "continues to signify just that very same thing forever." In a word, it's dead.

Thinkers have been analyzing and debating this passage for ages, because Socrates got it so wrong. His reaction to writing is typical of the confusion and anxiety new technologies often cause. Like the Luddites of today who believe that digital technologies are irredeemably inferior to older devices and even dangerous, he judged the new tool exclusively through the lens of the old one. Because writing didn't work just like conversation, he felt, it couldn't possibly be worth much and would only make people dumber. To Socrates, writing was useful only as an aid to oral dialogue, a kind of script, which is exactly how he and Phaedrus employ it.

What led Socrates to this narrow, pessimistic view of writing? He failed to understand that new connective technologies come along to solve genuine problems, and those problems usually have something to do with distance. In primitive times, the problem had been *psychic* distance; people were trapped in their own thoughts without an effective way to express themselves. Conversation solved this problem by allowing them to put their thoughts into words that could be shared and understood.

Oral communication was a great success, but it gave rise to a new problem of *physical* distance, rooted in the fact that conversation could happen only in close proximity to others. As civilization expanded, it became increasingly useful and important for people to communicate across great distances. By the fifth century B.C., merchants and traders were running businesses that spanned mountains, deserts, and seas. There were city-states and emerging empires whose political and military leaders needed to send messages to far-flung locations. Human messengers long met this need, delivering information by voice. But this system had drawbacks, including the limitations of memory. Written language solved the problem of physical distance by allowing words and ideas to travel anywhere and arrive intact, exactly as originally recorded. Writing also solved the *temporal* problem of storage, making it possible for information to be stored over the long term more reliably than it could ever be stored in the human mind.

As Plato shows in *Phaedrus*, this immensely practical innovation also had a less tangible, but ultimately far more significant, benefit. It allowed individuals to experience other people and their ideas *at a distance*, in a private, reflective way. A text written in a busy city could be "replayed" anywhere, including on the bank of a gurgling stream. Immediately after Phaedrus removes the scroll from his cloak, the two men step into the stream, which Phaedrus observes is "lovely, pure and clear"— a metaphor, perhaps, for what's about to happen to the flow of their thoughts. While closing one kind of distance, written language opened another, giving the mind a new kind of freedom. As a result of that freedom, writing turned out to be much more than a static record of old thoughts. Over time, it would become the fantastic medium for exchanging ideas and growing new ones that it is today.

Given who Socrates was, a philosopher whose life's work was embedded in the old medium, it's understandable that he didn't grasp the value of the new one. Steeped in the culture of the voice, he never imagined that one could go off alone with a written text and read it silently and thereby gain new insights. His doubts may also have been related to the physicality of writing. A firm believer that the mind was the source of all meaning, he was suspicious of the body and, indeed, the entire physical world. At one point in this dialogue, he refers to the body disparagingly as a mere shell for the intellect, "this thing we are carrying around." To him, a written text was just another "thing," a dumb object that pretended to do what the mind does but never could.

Plato had more vision than his teacher about the value of distance. As the action of the dialogue shows, he understood that there was much to be gained by retreating physically from the crowd. Years after Socrates' death, when Plato decided to open his own school, he founded it outside Athens in the same kind of countryside where this dialogue takes place. The Platonic Academy would become synonymous with the best of Greek thought, further evidence that there really is something divine about distance.

Second, though there's no record of what Plato personally thought about written language, he left plenty of evidence that he thought better of it than Socrates did. Plato also took a dim view of physical objects as sources of wisdom, but that didn't stop him from putting pen to scroll and becoming a writer himself. The reason we're able to read this dialogue today is that Plato wrote it down, using the very tool Socrates denounced. He was roughly forty years younger than Socrates and evidently more open to the possibilities of the new device. By recording in hard copy Socrates' dark fears about writing, he was effectively saying, "Sorry, old man, there's more to it than that."

For our purposes, in *Phaedrus* Plato establishes a basic principle on which to build a new way of thinking about digital connectedness: In a busy world, the path to depth and fulfillment begins with distance. The technological landscape is a great deal more complicated today, and over the centuries distance has taken on different meanings. But the basic dynamic hasn't changed: to steer your chariot toward a good life, it's essential to open some gaps between yourself and all the other chariots crashing around this busy world.

Technology is unpredictable, and the gaps often appear in surprising places. So far, digital gadgets have increased the general level of our busyness, creating a new need for distance. It's a problem yet to be solved, and it's worth noting that some 2,400 years ago, it was just beginning to dawn on people that they could use *their* newest technology for the opposite purpose: to reduce or temper their busyness. Might we be able to pull off the same trick in the digital age?

For that to happen, it's essential to be more mindful of how today's devices change our relationship to the crowd, which in turn affects our busyness and state of mind. Human connectedness is fluid and ever changing. When they first meet in the city, Socrates and Phaedrus are in a busy, highly connected situation. By talking a walk, they become less connected to the crowd and more connected to each other—and the scroll helps make it all happen.

As new technologies are added to the mix, the permutations

and subtleties multiply. In Athens, the city was synonymous with the crowd. But today, walking down a bustling city street can be a form of *dis*connectedness from the crowd, especially if you've just come from an office crowded with screens. While you're walking down that city street, if your mobile buzzes with a call or message, your relationship to the crowd changes yet again.

To make sense of all this, it's helpful to imagine connectedness as a continuum along which we're moving all the time. It's pictured below as a straight line between two poles, which I've labeled with the Greek letters alpha and omega. Alpha represents minimum connectedness, or the self alone, while omega is the maximum connectedness of the crowd.

The poles represent not just the fact of being in a crowd or being alone but the types of experience associated with those situations. When we're alone, our thoughts and feelings are oriented inward, and experience tends to be relatively quiet and slow. In contrast, in a crowd—whether physical or virtual— our orientation is more external, simply because there's more happening, more demands on our attention. Life in a crowd is typically busier and faster.

The rest of the continuum represents the range of situations between these extremes. Moving from left to right, solitude gives way to interaction with others, and one's experience

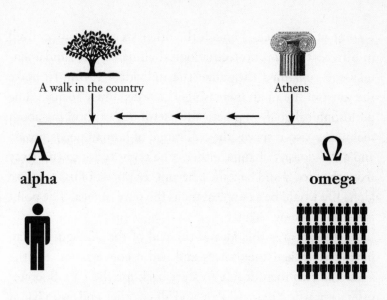

A walk in the country   Athens

A
alpha

Ω
omega

becomes relatively more outward and busy. Moving from right to left, the crowd grows smaller, and experience is relatively less busy and more inward. When Socrates and Phaedrus leave the city, they dramatically reduce the intensity of their connectedness, shifting from the omega end of the continuum toward alpha. Distance makes all the difference.

This is just a simple graphic device, and it can't begin to represent the full range of human experience. Everyone's temperament is unique, and we all have our own personal reactions to crowds as well as to solitude. There are born introverts as well as extroverts, and countless shades in between. A situation that feels oppressively crowded and busy to you might not strike me in the same way. Still, there is a rough correlation between how immersed anyone is in a crowd and how busy (or not) their thoughts are. And this idea is central to understanding the workings of human connectedness.

In the chapters to come, as the story progresses from Plato's era to the present, I'll occasionally use this continuum as a point of reference. Though the other six philosophers lived in different times and technological climates, the fundamental issue remained the same: the individual trying to make the most of life in an increasingly crowded, busy society. The philosophical goal—a practically useful way of thinking about technology, so it serves the full range of human needs, inside and out—doesn't change, either. The point is not to run away from the crowd and become a hermit. For most of us, the pure alpha life would be as unpleasant as the pure omega. The point is to find a happy balance.

Plato captures this idea at the end of the dialogue, when, having refreshed themselves and had a conversation for the ages, the two men decide to start back for the city. Socrates offers a prayer: "Beloved Pan, and all ye other gods who haunt this place, give me beauty in the inward soul; and may the outward and inward man be at one."

Chapter Six

THE SPA OF THE MIND

Seneca on Inner Space

"I force my mind to become self-absorbed and not let outside things distract it. There can be absolute bedlam without so long as there is no commotion within."

Not long ago, I began using my computer in a new way. Late at night, when the dishes were done and the bedtime stories over, I'd hole up in my home office for a half hour and watch music videos. I'm into jazz from the late 1950s and early '60s, a period one might not expect to be well represented online. Yet there's a rich trove of vintage clips out there from old television shows, films, and other sources, which ardent fans have taken the trouble to find and upload. Search for "Coltrane" or "Nina Simone" on the popular site YouTube, and within seconds you're watching a drop-dead performance you never knew existed until that moment, thanks to the enthusiasm and industry of someone identified only as DavidB87523 or NinaFreek.

Though these private concerts took place in what was technically leisure time, their purpose was broader than fun and diversion. This was a way of clearing my head at the end of a hectic workday, a personal version of the walk outside the walls of Athens recounted by Plato. No matter what kind of work one does, it's essential to step away on a regular basis, to recharge and gain perspective. In an always-connected world, the need for these gaps is more urgent than ever, yet harder to find.

With wireless signals nearly ubiquitous, physical escape of the Greek sort is practically impossible. So it's become necessary to create gaps *within* one's connected life. This is why some businesses and other organizations have begun asking, and in some cases demanding, that employees *not* check their office e-mail on weekends. All work and no play makes Jack a dull member of the team.

When you work for yourself, you have to be your own mental watchdog. Nothing wakes up my sense of inner freedom, or approximates the "divine" possession described by Socrates, like a great jazz performance. And now there was a whole new way to enjoy that experience, on the same screens where we do our work and run the rest of our lives. It seemed an ideal reprieve from the grind, a digital answer to a digital dilemma. Except for one problem: I was indeed using the *same screen* where so many other things were happening, in my world and everyone else's. While I had the crowd to thank for these videos, I also had to contend with the crowd as I watched them. And this turned out to be a challenge.

I've enjoyed jazz in many different situations, including live shows at clubs, concert halls, and outdoor festivals. I've listened to recordings at home, in the car, and on my iPod and occasionally watched performances on television or a movie screen. So I had a considerable database of experience with which to compare these computer-screen encounters. There's no question they brought me pleasure and occasionally left me gaping. Watching Miles Davis is qualitatively different from just hearing him; my eyes confirm that it *was* just a man, not a god, who produced those sounds.

Yet the video sessions fell short of the stroll outside the city walls I'd hoped for, because they were playing on a screen hooked up live to the digital grid, with its never-ending buzz of distractions. And that made it a lot harder to put any real distance between myself and the crowd. Instead of wandering down to the stream, I was stranded in the busiest city ever built, the digital one.

One night, for example, I was playing a clip from the Newport Jazz Festival of 1958. Dinah Washington was singing "All of Me" and she was on fire, thrilling to watch. *All of me. Why not take aaaaaaaall of me?* But the video is just a part of the experience on sites such as YouTube, which is designed to play up its participatory, user-driven approach. Using the site, you're constantly reminded that you're in the midst of millions of others whose presence affects everything you see and hear.

The page informed me that the video had been viewed more than 100,000 times, and rated by 203 people, with an impressive average rating of five stars. There were dozens of comments, many of which had multiple replies. One commenter noted that the singer "is my second cousin who I never met but I can't get enough of her music." I could feel all sorts of people hovering around the edges of the screen, jabbering away and inviting me to join in. To help me do this, there were buttons for sharing the clip, making it a favorite, flagging it, and adding it to my playlists. There were thumbs-ups and thumbs-downs and an invitation to post a video response. And there were direct links to the most popular social networks, where I could alert my friends to this find.

To the right of the box in which the video was playing was an advertisement, part of which unexpectedly—yikes!—expanded when I moved my mouse across it. Bannered across the top was a notice that YouTube would soon be "phasing out support" for my browser and I should upgrade to a "more modern" one now.

Digital technology is configured to encourage busyness, but then, so are we. As I took all this in while intermittently watching Dinah, my right index finger rested lightly on the clicker, trained by years of grazing to layer on the tasks and stimuli by seeking something new every few moments. There were other Web pages and applications already open behind and beside this one. Though I wasn't there for any of that stuff—this was supposedly my own personal after-hours jazz club, a place to sit back and do one delicious thing—the mind is adept at rationalizing its wandering eye. Isn't curiosity a virtue? Why *not* check my inbox? What were my friends doing right now over on social network X? What had happened in the world since I'd last scanned the news at media site Y?

I gave in to the last of these urges. "Enormous Jellyfish Sink Japanese Fishing Boat," said the first headline to come up, with a color photo of the dreadful monster. Shocking! "1 Dead, Others Reportedly Shot at Oregon Office Park," "Threat of Satellite Collision Grows," said one of the sidebar items, which, though disturbing, somehow hadn't made the nearby list of Most Read Stories, which I naturally checked out, top to bottom.

Now that my attention was loosed, there was no turning back. I clicked over to one of my e-mail services, which brought yet another menu of news headlines including, at last, something positive: "Ex–MLB Pitcher's Mom Rescued from Kidnappers." I didn't even know she was gone.

None of these side trips served any useful purpose. There was no reason for me to throw a plague of giant Pacific jellyfish into my cognitive-emotional stew, not at this moment. I was just yielding to the momentum of my own busy mind and thereby undermining the task at hand, which was to make it *less* busy, less full of the outward world and everyone in it.

I like other human beings quite a lot. I love my family and friends. I feel very lucky to live in a small town where I can walk down Main Street and run into half a dozen nice people I know. I relish the thought that there are billions of other people out there I don't know but one day might. That we can all now share our thoughts and interests with one another so easily, without meeting in person, strikes me as a tremendous step forward for humanity. The fact that I can type a couple of words and be transported instantly to a fabulous musical moment from a decade when I wasn't even alive, thanks to the kindness of a stranger—it's just beautiful.

Having been transported to that moment, however, I want to make the most of it. I want to experience what that jazz-loving stranger gave me in the way it deserves to be experienced. I want to be *in* the moments the music and images are creating, to know them in all their richness. For that to happen, I need to get away from the digital crowd and the claims it makes on my consciousness. Not completely away, because then I wouldn't be able to enjoy Dinah. Just far enough that I don't feel that caffeinated click-click-click of the mind. But the way screen life works, it's extremely hard to do so.

Somewhere in the background, Dinah was still at it. *Come on, baby, come on, Daddy, and get ALL of meeee!* She sounded as awesome as ever, but most of *me* had moved on. I'd broken the spell, pulled away from a source of intense pleasure and release without fully realizing I was doing so or knowing why. That wasn't the jazz experience I'd had in mind and knew so well from the past. It was *jazzus interruptus*, an unsatisfying pantomime of the real thing. Rather than shaking off the jumbled, vaguely unfulfilled state of consciousness my work-day had left me in, I'd intensified it. This effort to free my inner self of its burdens had landed me in a jail without bars, the restless, outward-leaning screen state of mind—which, weirdly, stays with you long after you step away from the screen. Some nights, I swear, it follows me not just to bed but into my dreams.

In a sense, I was holding those videos to an unfair standard. YouTube doesn't pretend to be a tranquil grove. If I really wanted a pure music immersion, sans distractions, I could have taken six steps into the living room and fired up the old stereo. However, the point of this exercise was to take advantage of something potentially wonderful about the *new* technology, a way in which the key source of my busyness might also serve as its antidote.

This is an increasingly important question, as the digital zone becomes the destination for more of *everything* we do. Work, family, friendships, thinking, reading—so much of life is migrating to these machines with their ever-expanding universe of information and potential tasks. Technology companies tout the many-splendored connectedness of digital tools as their chief advantage: the more people and information you can connect with and the faster and more intensely, the better. But after a while, all that flitting around does something terrible to your inner life. It denies you the very thing you went to the screen for in the first place: happiness.

My failed jazz experiment was just a small specimen of this paradox, a minor disappointment to me and of no consequence to anyone else. But its smallness is what makes it so telling. If I can't use this new medium to open up a tiny gap between myself and the crowd, long after the workday has ended, at an hour that's all about quiet and retreat, how is anyone going to do it while attending to the truly urgent needs of jobs, relationships, and other important aspects of life? The effort I was making to clear some space for my thoughts is emblematic of a much larger struggle with momentous implications. The frazzled, overconnected inner self is reporting to work every day in offices around the world. It's running corporations, universities, and nations. It's raising children and teaching them in school. It's trying to compose symphonies and write novels. It's working to fight hunger and poverty, end wars, find cures for disease. And it's trying to do all that while simultaneously navigating the ubiquitous digital crowd.

If distance is as valuable as Plato suggests, where on earth are we going to find it?

This would seem to be an entirely new dilemma, a lamentable by-product of the latest technologies. It certainly *feels* like a recent phenomenon. In fact, it isn't new at all. As the world has grown steadily more connected over the centuries, the physical distance Socrates and Phaedrus enjoyed has been losing its power to relieve the overloaded self. Even in ancient societies, people found it hard to escape their own busyness. The burdens and distractions of the city had a way of following them everywhere. The mind of two thousand years ago often felt hounded, too, cornered, with no place to hide. And back then, as now, there was a need for creative solutions.

One of the first thinkers to recognize this was the Roman philosopher Lucius Annaeus Seneca. He was born around the same time as Christ in the Spanish city of Córdoba, an important outpost of the Roman Empire, where his father was an imperial official. At a very young age he was sent to Rome for his education, and he went on to pursue a career in the government. By his midthirties he was a senator, and he later became a senior adviser to Nero during the latter's first eight years as emperor. Nero was just a teenager when he took the throne, and Seneca wound up effectively running the empire during those years, wielding such power that one twentieth-century historian called him "the real master of the world." Scholars have rated this period among the empire's best, and Seneca is widely credited for its success.*

But it was as a philosopher that Seneca made his most lasting contribution. He left behind a remarkable body of writings in which he wrestles with what constitutes a good, happy life and how to find it. He was a very busy man writing from the thick of life, and he knew how to get to the point and say it memorably. Writing about wealth and poverty, for instance, he says, "It is not the man who has too little who is poor, but the one who hankers after more." He believed the primary mission of philosophy should be to offer people practical advice on how to live better. His essays and letters often feel as if they were written not 2,000 years ago but last week. This is particularly the case when he's talking about the problem of life in a crowded world and the difficulty of finding a space apart for the mind. Mastering the *outer* world is one thing, but it's an even harder trick, he realized, to master the *inner* one, especially when you live at a time when the two are at odds. We live in such a time, and so did Seneca.

The Romans had taken the Greek concept of civilization, which had reached its highest expression in Athens, and expanded it to dramatic new proportions. Their empire stretched from northern Africa to Britain and east across Europe all the way to the Middle East. They devised various ingenious ways of pulling this vast domain together, to make it more efficient and manageable, including an excellent system of roads, a far-flung but highly disciplined army and civil service, and an extensive postal system. In addition to making Rome run more smoothly, all of this had the effect of shrinking the distances between places and people.

---

* *Nero's descent into corruption and depravity came later, after Seneca had been removed from power.*

Rome represented a new kind of connectedness, one that offered tremendous benefits, especially to the privileged classes, while simultaneously exacting costs, some of them quite high. In making the world smaller, the empire increased the everyday busyness and burdens of the individual. Life in the city of Rome itself was fast-paced, crowded, noisy. A weary soul could certainly escape, as the wealthy often did by retreating to the lavish estates they built in the countryside beyond the city. But even out there, they hadn't really left Rome, not truly. No matter where you went in the empire, there were frequent reminders—the roads, aqueducts, and fortifications, the legionnaires and postal carriers—that you were still *inside* a political, social, and cultural system that, in various subtle and not-so-subtle ways, demanded a great deal of you in terms of time, energy, and personal autonomy.

Adding further to the busyness was written communication, a technology that had taken off in the four hundred years since Plato. Writing had transformed life in the Mediterranean world and was a crucial factor in the success of Rome. The legal and administrative machinery that held the empire together depended on written laws, edicts, records, and communiqués. The age of paperwork had arrived (though at this point it was papyrus-work). Writing also figured hugely in the everyday lives of literate Romans such as Seneca. Postal deliveries were important events, as urgently monitored as e-mail is today. Seneca writes at one point of his neighbors hurrying "from all directions" to meet the latest mail boat from Egypt. Books were now central to education, and literacy in both Greek and Latin was essential for any Roman seeking to rise to a high position in society. In a world increasingly driven and defined by written language, there was just a great deal more information to process and absorb.

In sum, the busy Roman was constantly navigating

crowds—not just the physical ones that filled the streets and amphitheaters but the virtual crowd of the larger empire and the torrents of information it produced. Seneca spent most of his life at the throbbing center of it all. Though he flourished in the crowd, he also struggled with its demands and was acutely conscious that if he were not careful, it could take over his life. The best record of his thoughts on this subject is a series of 124 letters he wrote to an old friend named Lucilius, a career civil servant who was deeply interested in philosophy and apparently viewed Seneca as a teacher and role model.

In his youth, Seneca had embraced Stoicism, a Greek school of thought that emphasized self-reliance and simple living. Today the word "stoic" has a dour, joyless ring, but the philosophy itself is positive and life-embracing, particularly in the hands of the unfailingly upbeat Seneca. His correspondence with Lucilius, known as the *Epistulae Morales* or "moral letters," covers an incredibly broad range of topics, from the mundane (a virus he's just caught) to the transcendent (why death shouldn't scare us), sometimes in the same sentence. One of his most frequent themes is the danger of allowing others—not just friends and colleagues but the masses—to exert too much influence on one's thinking. The more connected a society gets, the easier it is to become a creature of that connectedness. One's inner life grows increasingly contingent, defined by what others say and do. "You ask me to say what you should consider it particularly important to avoid," begins one letter. "My answer is this: a mass crowd. It is something to which you cannot entrust yourself yet without risk. . . . I never come back home with quite the same moral character I went out with; something or other becomes unsettled where I had achieved internal peace."

To fend off the crowd, Stoics believed, it was essential to cultivate inner self-sufficiency, and Seneca returns to this notion over and over. Learn to be content within yourself, to trust your own instincts and ideas. Those who achieve this autonomy, he argues, are best able to enjoy and make the most of their outward lives. They thrive in the crowd because they're not dependent on it.

This was hard to do in a society that placed as many demands on the individual as Rome did. The movers and shakers in Seneca's circle were terribly busy, constantly rushing about with what he called "the restless energy of a hunted mind." He paid particular attention to two aspects of this restlessness. One was the ceaseless need to travel, as if happiness always lay off in some distant city or resort. Those who lived this way were really just running away from themselves and their worries, Seneca said. And they were bound to fail because the stressed-out mind had a way of carrying its burdens everywhere. "All this hurrying from place to place won't bring you any relief, for you're traveling in the company of your own emotions, followed by your troubles all the way."

He noticed that those who lead distracted, unsettled lives will go to great lengths to *remain* that way, even on vacation. "The man who spends his time choosing one resort after another in a hunt for peace and quiet, will in every place he visits find something to prevent him from relaxing." In other words, by the first century A.D. the busy, crowd-induced state of mind had gone mobile, and even then it was hard to shake. Today we ask, "Does this hotel have Wi-Fi?"

The second variety of mindless bustling was the way people consumed *information*. Due to the explosion of writing, the empire was awash in texts. The book collection of the famous Alexandria library now numbered in the hundreds of thousands, but you didn't have to go to Egypt to read. A Roman bookseller with a large staff of scribes trained in dictation could churn out copies of popular books very quickly.

Plus there was the mail traffic, the paperwork that propelled both government and commerce, and other kinds of written communication. Elite, literate Romans were discovering the great paradox of information: the more of it that's available, the harder it is to be truly knowledgeable. It was impossible to process it all in a thoughtful way. So there was a tendency to graze, skim the surface, look for shortcuts.

Seneca observed that people had begun to read in the same way they traveled, racing harum-scarum from book to book. Some never took the time to develop an intimate familiarity with the ideas of a single great writer, a practice he'd found useful in his efforts to develop his own mind and beliefs. There was more to be gained from knowing one excellent thinker deeply, he believed, than from knowing dozens superficially. Instead, he wrote, readers "skip from one to another, paying flying visits to them all." Reading in this fashion is like anything else done in haste: "Food that is vomited up as soon as it is eaten is not assimilated into the body and does not do one any good . . . a plant which is frequently moved never grows strong. Nothing is so useful that it can be of any service in the mere passing. A multitude of books only gets in one's way."

He might as well having been writing in this century, when it's hard to think of anything that *isn't* done in "mere passing," and much of life is beginning to resemble a plant that never puts down roots.

There are essentially two ways to deal with this problem. One is to surrender to the madness, allow the crowd to lead you around by the nose and your experience to become ever more shallow. Seneca tells the story of a rich man named Sabinus who didn't read much but was desperate to *appear* as if he did. At great cost, he purchased about a dozen slaves and made them memorize the works of famous writers. One had to learn

Homer by heart, another was assigned Hesiod, and so on. At dinner parties, he would keep the slaves "at his elbow so that he could continually be turning to them for quotations from these poets," which he would then recite for his guests as if he'd produced them himself. Today, this kind of faux wisdom is a lot easier to obtain. With our screens Googling away, all the brilliance in the world is at our fingertips. And as long as it remains there, rather than in the mind, how different are we from Sabinus?

The other option is to step back, recognize you're too busy, and pare down. "Measure your life: it just does not have room for so much," Seneca advises Lucilius. Though much of what keeps us hopping is unavoidable—the demands of work and other inflexible obligations—a fair amount is pure, self-created bustle. *Why* check the inbox ten times an hour on a Saturday, when once will do? By eliminating the worthless time-wasting stuff and focusing on what serves your highest purposes, Seneca argued, you can shape and enrich your own experience. Even in his time, this was not a new revelation. It's the commonsense view behind Socrates' notion that each of us is holding the reins of our own inner chariot, deciding at every moment where and how to deploy our energies. As Winifred Gallagher writes in her recent book *Rapt: Attention and the Focused Life*, "It's about treating your mind as you would a private garden and being as careful as possible about what you introduce and allow to grow there."

"After running over a lot of different thoughts," Seneca tells Lucilius, "pick out one to be digested thoroughly that day. This is what I do myself; out of the many bits I have been reading I lay hold of one."

The question is how to apply this idea to a world as crowded with information and stimuli as ours. It's one thing to select an idea or experience you want to focus on, another to tune out all the *other* stuff around it. What do you do when even Dinah Washington can't hold your attention?

In one of the letters, Seneca elaborates further on his technique. "I cannot for the life of me see that quiet is as necessary to a person who has shut himself away to do some studying as it is usually thought to be," he begins, noting that he's writing from a room directly above a public bathhouse or spa. There were facilities like this in every Roman city, and they figured prominently in day-to-day life. Romans went to the baths not just to bathe but for exercise, massages, and other body treatments, as well as socializing and simple relaxation. They were the progenitors of today's spas and, like the latter, not always the tranquil oases we like to imagine. "Picture me with a babel of noise going on all about me," writes Seneca, describing in great detail the sounds rising up from below, the grunting and gasping of the weight lifters, the smack of hands on the shoulders of someone getting a massage, the raucous shouts of people leaping into the pool "with a tremendous splash." As if this weren't enough, street sounds are drifting in, too, from vendors hawking drinks, sausages, and other edibles, from "the carriages hurrying by in the street, the carpenter who works in the same block, a man in the neighborhood who saws, and this fellow tuning horns and flutes at the Trickling Fountain and emitting blasts instead of music."

A crazy scene, yet he tells Lucilius it's not bothering him a bit. "I swear I no more notice all this roar of noise than I do the sound of waves or falling water." How can that be? He explains that he's trained himself to be oblivious. "I force my mind to become self-absorbed and not let outside things distract it. There can be absolute bedlam without so long as there is no commotion within."

He makes it sound easy and completely self-driven, as if there's a switch he can turn on and off in his mind. The mind, he writes in another letter, should be "able at will to provide its own seclusion even in crowded moments." Perhaps, but in achieving what he calls "inward detachment," he's also had a crucial assist from a tool: the letter he's writing to Lucilius. He doesn't give it credit, but there's no question that the act of writing has helped him focus his thoughts. The letter is the *object* of the journey inward, and it works beautifully, taking him to his desired destination. This letter is one of his liveliest and most interesting, which surely wouldn't have been the case if he'd been distracted by the ambient racket. In fact, toward the end of the letter he reveals that he wrote it as an experiment, in a conscious effort to see if he could filter out the din.

What's remarkable is that he's using a technology that played a large role in making the Roman mind busier—written language—to reduce that busyness. For a man who lived by a philosophy of simplicity and inner autonomy, letter writing was a perfect remedy for the problem he faced in that noisy room. First, it simplified the crowd by reducing it to one person. Of all the people in the gargantuan throng that was Rome, he chose one as the exclusive focus of his thoughts. Lucilius is the human equivalent of the one thought he would choose each day for special attention. Second, by muffling the distractions of the outward world, letter writing allowed him go back inward and reclaim his autonomy.

Thus, it turns out that written language had the capacity to do exactly what Socrates said it never could: set the mind free, so it could do its best work. It did this by affording a private, reflective experience. What Seneca describes resembles the state of "flow" that the modern psychologist Mihaly Csikszentmihalyi has identified as the best kind of human experience. In essence, flow is what happens when one is so absorbed in an activity that the world seems to fall away. The activity can be as simple as working on a jigsaw puzzle or as complicated as flying a plane, as long as it produces what he calls "a deep but effortless involvement that removes from awareness the worries and frustrations of everyday life." In flow, there is no sense of time or distraction, just complete immersion in the moment. According to Csikszentmihalyi, one achieves this happy state by learning to "control inner experience" and find "order in consciousness." The pursuits that induce it tend to have a sense of boundedness or limits; most are goal-directed tasks that have a reasonable chance of being completed. There's no satisfaction in doing a puzzle that doesn't fit together or shooting baskets against a backboard with no hoop.

Writing a letter meets all these criteria. In a society where physical distance was no longer an effective escape from the crowd, Seneca sat down with a blank page and escaped in another way: he found *inner* distance.

Written language didn't always work this miracle, of course. Like a book read too hurriedly, a letter dashed off without much thought would be more of a "flying visit" and probably not much help. It's obvious from reading Seneca's letters that he gave himself fully to the writing of them. This is why they've proven useful to so many readers and survived for so long. More than 1,500 years after his death, Queen Elizabeth I used to start her day with his advice on how to quiet the mind. She would often translate Seneca from Latin into English, thus engaging in the very same activity that had worked for him in the room over the spa, writing.

Could it work for us? The goal of my online jazz sessions had similarly been to turn the technology of the present back on itself. I didn't have much success, but perhaps that was because I hadn't used the Senecan approach. That is, I hadn't viewed my situation as the *philosophical* problem it was. As a busy man in a busy society, Seneca often found himself stuck at the far end of the continuum of connectedness, the overcrowded place I call omega. So he developed a few practical techniques, grounded in his Stoic philosophy, for moving back toward alpha. Choose one idea a day to think about more deeply. Train the mind to tune out the chaos, through the art of concentration. These may not be startling insights, but the fact is, they hadn't occurred to *me*. Seneca was more conscious and thoughtful about his overload problem than I'd been about mine.

I went back to the screen and tried again with the same video. This time, before I even sat down, I contemplated my own craving for inwardness and what I might do differently this time that would help me attain it. I made sure there were no other distracting applications open on my screen. When I called up the video on YouTube, I immediately looked for a way of blocking out the user comments and all the other crowd-oriented bells and whistles that had been the start of my mind's wanderings.

There were a couple of options. One was a button that would allow me to view the video by itself in a new window, while leaving the original window (with all the bells and whistles) open and partially visible. Another would maximize the video so it took up the whole screen. The latter was more in tune with my aim—to maximize just one thing, minimize the rest—so I went for that. As soon as I did so, I remembered why I rarely used that button: there's a technological price to pay for enlarging a video; in chasing psychic seclusion, I lost clarity and resolution. Though I wouldn't want to view my videos that way all the time, I could live with it for the purposes of the experiment.

Since this time I was really trying to focus, I suppose it was inevitable that I would do better. In fact, I stayed with Dinah straight to the end, no side trips. When she was done, I closed the browser, stood up, turned off my desk lamp, and abandoned the screen. Walking across the house, I tried to determine if

I felt any different. Consciousness can't be measured precisely like body temperature, and it's hard to compare states of mind across time. The music and images had certainly been terrific, and by not darting around I felt I'd enjoyed it all more fully. Still, I couldn't honestly say that there had been, in Seneca's words, "no commotion within."

Though the full-screen view had been less busy and I'd restrained myself from veering off, this time something else had gotten into the way. Purely because the experience was still taking place on a digitally connected screen—the green modem and router lights were glowing off to my left the entire time—my mind was in connect mode from the moment I entered my office. It's almost Pavlovian: I see the screen, I know it's connected, and my thoughts shift into a different gear. I associate the tool so closely with outwardness that it's hard even to think of it as a tool of inwardness. And since the exercise was all about thinking, that was a problem.

In a sense, therefore, the issue was in my head. But the fact is, the technology itself doesn't help much. Seneca was lucky enough to have at hand a tool that's unusually good at inducing concentration. Today's machines could go a lot further toward helping us out. Having made the philosophical choice for a more inward life, I would have appreciated a device that acknowledged focus as a worthy goal and offered easy ways to achieve it. The various sorts of busyness I was forced to block out of the experience were manifestations of the maximalist bias of digital devices. Our screens are designed to keep us as connected and busy as possible, in full-on crowd mode. And they make it very difficult to shift out of it. Gadgets now exist that allow one to remove some of the more annoying distractions from the online experience, but they're add-ons, and they haven't changed the fundamental tendency.

It was late when my experiment ended, and I did something I don't normally do at bedtime. I took a portable radio into the bedroom and tuned in to a jazz show out of Boston that I usually listen to in the kitchen in the early evening. Somehow, the fact that it was a *local* radio station, arriving not through a digital connection but over the airwaves, was important. Lying there in the dark listening locally was a way of completing what I'd started with the video—putting some distance between myself and the big, busy, connected world.

Chapter Seven

LITTLE MIRRORS

Gutenberg and the Business of Inwardness

"Even before the books were finished, there were customers ready to buy them."

When one of the most anticipated gadgets of the last decade, the Apple iPhone 3G, arrived in stores around the world a few summers ago, it was hailed as a wonder. This was not just a phone but a sleek pocket computer that did it all: web browser, camera, video and music player, navigational device, and many other things. It could also run on faster networks than the original iPhone, allowing the user to perform more tasks more quickly. "Even better," enthused one magazine reviewer, "3G coverage enables you to make a phone call and surf the Web at the same time." In other words, by the currently prevailing philosophy of digital life—the more connected you are to more people and information at all times, the better—it was a dream machine. And it set off one of those mass consumer frenzies that have become a kind of global ceremony, with quasi-religious overtones.

"Lines for what the faithful call the 'Jesus phone' started forming early Friday outside Apple Stores from Silicon Valley to Hong Kong," reported the *Mercury News* of northern

California, "with hardy souls bringing sleeping bags, laptops and a desire to bond with fellow iPhone acolytes." At one Tokyo store, the line was half a mile long, the atmosphere pure pandemonium. "The store's entrance was besieged by reporters and camera crews," according to a news report, "while helicopters circled overhead as an LED display counted down to when the handset went on sale." The first customer in line, one Hiroyuki Sano, had traveled 220 miles and camped outside the store for three days. When he walked out with his prize, a large pack of tech paparazzi chased him for four blocks, trying to get a quote. "I'm extremely happy," the reportedly breathless Sano finally said.

And why not? He was holding the ultimate handheld. We're always happy when we purchase a shiny new connective device, because we're thinking about all the interesting, useful, fun things it will do for us. And how much better it will be than the old device, which was fine but didn't *quite* meet our needs. The new model is the answer, and it's worth braving the crowd to get it.

When we take it home and start using it, however, a problem arises: we're still in a crowd, the screen crowd that's every bit as crazy as the one in Tokyo. The very thing that makes these technologies so popular, their capacity to keep us in closer, more constant touch with the rest of the planet, is what makes them such a burden. They're better at making us busier. Their greatest strength is their greatest weakness. We're as hounded by our screens as poor Mr. Sano was by the paparazzi, except we invite this crowd to chase us. We carry it in our pockets and respond to its every demand.

If having an inward life matters as much as having an outward one, we've created a technological quandary for ourselves. The gadgets we buy and use every day are designed, built, and marketed on the premise that it's an unalloyed good to always be in the crowd. And that turns out to be a lousy idea. It makes no sense to work and live in this fractured, always-on-call fashion. Rather than bringing the crowd ever closer, our machines should help us find some distance, whenever we need it.

Based on recent experience, it's hard to see how that's going to happen. On one level, we're at the mercy of the big technology companies, which make enormous profits from tools that seek to increase the intensity of our connectedness, pushing us further and further toward the crowded omega end of the spectrum. And there would seem to be no good business argument for any of them to adopt a new approach. Why change course when people everywhere are lining up to buy your products?

But those lines point to the true source of the problem: us. The merchants of technology aren't forcing anyone to buy their machines. We've signed on to the notion that the best devices are those that deliver maximum connectedness, and we've endorsed it with our wallets. In effect, we're designing our technological future, striving to make life even busier and harder to navigate than it is now.

Are we stuck in the crowd forever?

Not if the past is any guide. New connective devices have always had an outward bias. That's why they come along in the first place, to help people reach outward. New technologies form new crowds. And the better they are at it, the more eagerly they're adopted. However, because they tend to increase the individual's exposure to the crowd and ramp up the busyness, they strain the mind and the spirit. Thus, it becomes essential to find escape hatches. As the previous two chapters showed, in the ancient world forward-thinking people discovered clever ways to do just that. In ancient Athens, physical distance did the trick for two friends who fled the city with the help of a scroll. In frenetic Rome, Seneca found inner distance in the act of writing a letter. In both cases, the connective technology that was making the world smaller, written alphabet-based language, played a key role. A tool of outwardness could foster inwardness, too.

However, there was one aspect of written language that remained highly outward and crowd-oriented: reading. In ancient Greece and Rome and for most of the Middle Ages, reading was not the private activity we know today. For more than a thousand years, most reading was done *aloud*, much as Phaedrus read to Socrates from a scroll. People would sit in libraries and monasteries with books open in front of them, reading audibly, using their voices. The kind of reading we take for granted—an individual sitting silently with a book, eyes moving across the lines—was rare. Silent reading was so rare that when someone engaged in it, others remarked on it as curious and even eccentric.

Saint Augustine, who lived in Italy in the late fourth century A.D., notes in his *Confessions* that the bishop of Milan (known today as Saint Ambrose) had an unusual habit: "When he read, his eyes scanned the page and his heart sought out the meaning, but his voice was silent and his tongue still." Reading in this fashion was, according to Augustine, the bishop's way of "refreshing" his mind. This story, vividly retold by Alberto Manguel in his 1996 book *A History of Reading*, is the earliest known instance in Western history of a person who habitually read to himself without speaking.

Reading was, as Manguel put it, an "oral skill." It was also a social one, in that it generally happened in the company of others. Rather than reading privately, people gathered to be read to in groups. There were various reasons for this. Most people couldn't read at all, so oral culture had lived on since the genesis of writing, as had the belief that spoken expression was the highest form of communication. Collective reading was also a product of economics. Books were made by hand and, for most people, including many who could read, prohibitively expensive. After the fall of Rome, the Church was the producer and proprietor of most written information in Europe. Unless you were very rich and could afford your own library, it was only at Mass and other religious gatherings that most people came within reach of the beautiful hand-lettered tomes laboriously produced by Church scribes. In short, reading had not yet become the private, inward activity it is today.

During medieval times, this began to change. Some who had access to books started to read silently in private. And they discovered what a different experience it was from reading aloud and among others. To read privately without vocalizing the words was to take an inner journey, shared by no one else and not subject to outside influence or control. The mind was not simply "refreshed," as Augustine had observed, but liberated. Private reading allowed one to own the text in a new way. And having taken ownership, readers could wander *outside* the confines of the text, generating new thoughts and ideas for themselves. Certainly, this kind of thinking could also occur when books were read out loud, but because private reading was inherently inward, it was more conducive to it. For the pioneers of this practice, many of them scholarly monks, this was a revelation and no doubt exhilarating. It was a whole new way to enjoy the inner distance that Seneca had experienced while writing a letter.

Still, even as inward reading was embraced by the lucky few, for the many, books remained largely out of reach. Through the early fifteenth century, reading remained the crowd-oriented experience it had always been. And there was no reason to think that would ever change. First, there were serious technological constraints to making books more accessible. They were made of costly materials (animal skins, paper, ink) by skilled craftsmen who could work only so quickly. Second, the most powerful elements of society, the Church and the aristocracy, were not eager to promote wider availability of books and the inward experience they offered. The Church in particular understood that reading could be a route to unorthodox, heretical thoughts and therefore a threat. It was around the same time that silent reading became the norm among the educated, roughly the year 1000, that the first heretics were burned at the stake. There were enough troublemakers. Why encourage the rabble to think for themselves?

The notion that these constraints might be surmounted and private, inward reading somehow become available broadly to all kinds of people was as unlikely as the prospect of digital devices helping *us* get away from the crowd seems today. Impossible! Yet it was about to happen. And the way it happened demonstrates not just that technologies can be rethought in unexpected ways but that those who solve this particular riddle can reap enormous rewards.

In 1432, the German city of Aachen hosted an event that epitomized how crowded medieval life often was. Aachen, one of Europe's great cathedral cities, was a major destination of Christian pilgrimages like the one portrayed in Geoffrey Chaucer's *Canterbury Tales*. Pilgrims went to Aachen because its magnificent cathedral housed some of the most sacred relics in all of Christendom, including what were reputed to be the swaddling clothes of the infant Jesus, the Virgin Mary's robes, and the cloth used to wrap the severed head of John the Baptist. These objects were widely thought to have miraculous powers, and they drew such huge crowds that in the late 1300s Church officials decided they had to limit access to them. Henceforth the relics would be shown to the public only once every seven years, and for just two weeks.

During those septennial events, Aachen was inundated with pilgrims. It happened in 1432, with thousands pouring into the city from near and far on foot, horseback, and donkey, in carts, wagons, and any other means they could find. Like Chaucer's pilgrims, they were a motley assemblage representing a wide range of classes and circumstances. But having reached the cathedral, they merged into one crowd, a heaving, shouting, sweaty mass of humanity, all trying to reach the same goal, the sacred relics. One year, the pressure of all those bodies caused a building to collapse on the crowd, leaving seventeen dead and a hundred injured.

What exactly were they seeking? Tradition held that the relics sent out invisible rays imbued with divine powers that could heal the sick and answer other prayers. The surest way to obtain these blessings was to touch the objects. That had once been easy to do, but as the crowds had grown over the years it had become impossible to provide physical access to everyone—too many people, too little time. However, if you stood in the path of the rays and they hit you, it was thought to be just as good. Thus, during pilgrimages, the relics were moved to a raised platform outside the cathedral, where clerics held them aloft one at a time in order to give the rays wide distribution.

A special device had been created to ensure that no one missed out: a small convex mirror designed to catch and absorb the rays. The mirrors were fashioned from metal, often embellished with decorative designs, and sometimes worn as a badge. According to historian John Man, a pilgrim, having bought one from a local vendor, would find a vantage point with a straight line to the relics—some scaled the city walls—and hold it up "as if it were a third eye." Because the mirrors were thought to retain the sacred energy, they could be used long afterward to heal the blind, the sick, and anyone else in need of divine help. "You could head for home in the secure and happy knowledge that you carried in your belt-pouch the very essence of the miraculous," writes Man.

They were a handheld, mobile version of the relics, with excellent storage capacity, and, to those who believed in their power, they performed a valuable service. During the 1432 pilgrimage the mirrors were so popular that the local artisans who made them under a guild-controlled monopoly couldn't keep up with the demand. So it was decided that for future pilgrimages, craftsmen from other towns would be allowed to make and sell them, too. Since the gatherings could draw well over 100,000 people, it was a business opportunity with obvious potential. A new supply of mirrors would, as another historian puts it, "no doubt command a ready sale among the multitude."

One of those who jumped at it was Johann Gutenberg, an ambitious entrepreneur from Strasbourg, who had an original idea. Up to this point, the mirrors had been made by hand, a time-consuming task. Gutenberg believed there was a way to mass-produce them with an existing technology, the presses that had been used for centuries to make wine and olive oil. Rather than extracting liquid from fruit, a press could be used to punch out mirrors from sheets of metal, with each sheet yielding many mirrors. If it worked, Gutenberg would be making mirrors at a rate no artisan could match and at a much lower cost—a classic economies-of-scale strategy.

He brought some background to the task, having grown up around metalsmiths and coin makers, and he found three backers. They set up a partnership to manufacture mirrors for the next pilgrimage. Their plan was to sell 32,000 of them, with Gutenberg getting 50 percent of the profit while the other three split the remaining half. The mirrors were apparently made and sold, but there are no surviving records of how many, or whether Gutenberg made a profit on the venture.

For our purposes, however, what matters is the fundamental problem he was trying to solve and where it led him next. Gutenberg saw that there was money to be made in helping those of his time deal with the challenge of crowds. The pilgrims had arrived in Aachen en masse and willingly formed a crowd in order to obtain the blessings of the relics. It was an outward journey, but the goal was inward: to absorb the spiritual emanations of the relics into their bodies and souls and take them back home. The crowd presented an obstacle in two different ways: it was a physical obstacle blocking bodily access to the relics, and it was a virtual or effective obstacle, in that the sheer number of people made it impossible for local artisans to supply enough of the mirrors that surmounted the physical impediment by "connecting" through the air.

The horde outside the Aachen cathedral was different in countless ways from the digital crowd we find ourselves in lately. But the two crowds are alike in one important respect: we, too, are making outward journeys in pursuit of inward goals, using our "Jesus phones" to catch invisible signals. And the crowd is getting in *our* way, too. Why do you buy a new screen device in the first place? Is it merely to project yourself outward into the world, to connect for connecting's sake? No, you connect in order to do your work well, to communicate with friends, to learn and explore. Using a screen, you pursue those goals outwardly, by navigating the digital crowd. However, at the end of the day, what matters is what you take home. It's about the quality of your life experience, which is a product of how well the device helped you perform your outward tasks *while reaping inner rewards.*

Does your screen time help you think and work better? Does it deepen your ties to your friends? Does it help you find that much-needed distance and space? Do your explorations enrich your understanding of the world? Do you come away in a better state of mind than you were in to begin with? These are all interior questions. And the more time you spend in the digital crowd, the harder it is to answer them in the affirmative. Inner life becomes not deeper and happier but shallower and more unpleasant.

Gutenberg was a businessman and technologist, but his inventions had profound philosophical implications. In the case of the mirrors, mass production would allow more pilgrims to have their own private connection to the relics, from *within* the crowd. Amid all the noise and jostling around them, they could reach outward—in effect, leap over the crowd—to their goal and take it back inward. Strangely, today we're moving in the opposite direction, "upgrading" our digital gadgets in ways that make life more crowded rather than less, so that it's that much harder to open up gaps between oneself and the frantic world. At its worst, a digital screen becomes the opposite of the little mirrors, a talisman of bad energy, a portable curse.

Though the mirrors are just a metaphor for today's predicament, they led Gutenberg to something much bigger and more directly relevant to the present. In his next venture, he took the same technological principles—a press, mass production—and applied them to written communication, which, as we've seen, had its own crowd problem. Just as there hadn't been enough mirrors to satisfy popular demand, there weren't enough books to go around. As a result, most Europeans didn't have access to the inward experience that reading offered. Gutenberg rethought the technology of book production, developing a printing press with movable type that would allow books to be made more quickly and cheaply than they were made by hand. With his first printed text, a Bible with forty-two lines per page, he gave the world something wholly new: a machine-made book with a uniform text that could be reproduced with unprecedented efficiency. Over time, this would allow far more people to read by themselves in the private way that fosters inwardness.

It was an immediate, smashing success. In 1455, some pages from Gutenberg's first run of Bibles were shown at a trade fair in Frankfurt. A man who saw them named Enea Silvio Piccolomini (later Pope Pius II) wrote a letter to a high-ranking cleric, reporting that this new kind of book was remarkably easy to use, seemingly flawless. But he wasn't optimistic about getting his hands on one, because they were already a hit. "I shall try, as far as possible, to have one of these Bibles delivered for sale and I will purchase one copy for you," he wrote. "But I am afraid that this may not be possible, both because of the distance and because, so they say, even before the books were finished, there were customers ready to buy them."

Not everyone was thrilled with Gutenberg's creation. As today, there were pessimists and scolds who viewed new technology as a blight on civilization. In his recent book, *The Case for Books: Past, Present, and Future*, Robert Darnton quotes from a letter written in 1471 by an Italian scholar named Niccolò Perotti. Though he'd initially seen the printed book as a good thing, just a decade and a half into the print age, Perotti concluded it was a menace:

I see that things turned out quite differently from what I had hoped. Because now that anyone is free to print whatever they wish, they often disregard that which is best and instead write, merely for the sake of entertainment, what would best be forgotten, or, better still be *erased* from all books. And even when they write something worthwhile they twist it and corrupt it to the point where it would be much better to do without such books, rather than having a thousand copies spreading falsehoods over the whole world.

Like all great innovations, print spread quickly. As of 1480, there were presses operating in more than 120 European cities and towns. By 1500, this first wave of printers had churned out an estimated thirty thousand different titles and millions of copies. After centuries in which reading had been a largely outward, crowd-focused activity, the idea of having a book that one could read alone, in a completely interior fashion, turned out to be extremely powerful. Gutenberg had tapped into a hunger that was broader and deeper than anyone had anticipated.

The desire to extend this experience more widely, and perfect it, produced yet more innovations. The earliest printed books were modeled on manuscripts, which meant they were large (ideal for displaying and reading to large gatherings), beautiful, and still quite expensive to produce. But printers soon realized that there was a need for cheaper, smaller books better suited to everyday people reading just for themselves. By the early 1500s, there were pocket-sized books with new typefaces that were easier on the eye, along with other reader-friendly innovations.

The effects of this sea change in written communication were not strictly private, however, not by a long shot. In 1517, a German monk with some unorthodox theological ideas nailed his objections to various church policies and practices to a church door in Wittenberg and set off the Protestant Reformation. Martin Luther had reached these views through his own reading and biblical scholarship, and over time print technology allowed his message to reach a wide audience of readers, who in turn would read of this dramatic challenge to ecclesiastical authority and decide where they stood. Gutenberg's device played a crucial role in breaking the Church's hold on power, and in the subsequent political and social changes that would shape the modern world. The values of freedom and equality that we cherish today took root through the spread of reading and the power it conferred on individuals to think for themselves.

Of course, the book has many other strengths that have made it such a useful tool, and help explain why it has endured so far into the computer age. "[I]t has proven to be a marvelous machine," Darnton writes,

> great for packaging information, convenient to thumb through, comfortable to curl up with, superb for storage, and remarkably resistant to damage. It does not need to be upgraded or downloaded, accessed or booted, plugged into circuits or extracted from webs. Its design makes it a delight to the eye. Its shape makes it a pleasure to hold in the hand.

Does any of this make Gutenberg a philosopher? Not in the traditional sense. Little is known about the man himself, and there's no evidence that he consciously set out to democratize reading. He was first and foremost a businessman or, as John Man puts it, "an early capitalist" who recognized there was money to be made in mass-producing Bibles. Nonetheless, it took a philosophical mind to step back from the challenges of the crowd in late-medieval Europe and come up with not one but two very different solutions. If Gutenberg had thought only of mass-producing the mirrors, we wouldn't know his name today. If he'd thought only of the printing press, we would know his name but have less reason to surmise that he was consciously seeking solutions to the conundrum of the self in a crowded world. Because he had both ideas and worked at them so doggedly—he spent many years on the two ventures, borrowing money heavily along the way—it's quite possible that he was pursuing not just profits but a genuine vision.

And his work translates into ideas we can learn from today, when advertisements tout the "always-connected" life and influential technology critics praise new devices for the extent to which they broaden and speed up not just our reach but our availability, bringing the crowd ever closer in ever more dimensions of life.

When my trusty notebook computer picks up a wireless Internet signal, it tells me I'm "Connected!" and the exclamation point leaves no doubt what that means: Good news! I'm in the crowd! When there's no available signal, a red X appears over the wireless icon at the lower-right corner of my screen. There's no exclamation point now because this is nothing to be gleeful about. The crowd is gone. For years, I responded to the red X with frustration and impatience, sometimes a whispered curse and a hand slammed on the nearest hard surface. I could feel the pulse pounding in my neck as my blood pressure rose. I was a good maximalist, and there was nothing worse than losing the crowd.

I was wrong. I see that now, and I'm not the only one. It's clear that a full-time outward-focused life is unproductive, unhealthy, and unhappy in manifold ways. If you never lose the crowd, the magic never happens. We need distance and gaps, and we need them on a regular basis. Yet we press on, oblivious. Lately there's been an effort to make reading, the ultimate inward experience, more outward. Some e-reading devices allow you to toggle your attention back and forth between the text and the rest of the digital universe—the always-connected book. Enthusiasts of this approach predict that in the future all reading will be done effectively in public. That is, we'll be navigating links, comments, and real-time messages from distant others even as we try to read, say, a terrific novel. In a way, that would be a return to the pre-Gutenberg era, when the crowd looked askance at solitary, silent readers.

It's a very good thing to have broad access to information, which is why libraries have always been so valuable, and the bigger the library, the better. For research purposes, this Google age is a wonder. But there's a difference between *access* to information and the *experience* of it. Reading evolved away from the crowd for a reason: it wasn't the best way to read. Would anyone want to be trapped in a library in which all the books on all the shelves, and all the readers at all the desks, were talking out loud simultaneously? Hopping around among competing digital distractions, it's impossible to go truly inward, to become immersed in reading to the point where the crowd falls away, an experience poet William Stafford captured nicely in the lines

Closing the book, I find I have left my head
inside.

The point of the new reading technologies, it often seems, is to *avoid* deep immersion, precisely because it's an activity the crowd can't influence or control and thus a violation of the iron rule of digital existence: Never be alone. Deep, private reading and thought have begun to feel subversive. A decade ago, the digital space was heralded for the endless opportunities it offered for individual expression. The question now is how truly individual—as in bold, original, unique—you can be if you never step back from the crowd. When we think and write from within our busyness, surrounded by countless other voices, too often the result is reactive, derivative, short-shelf-life stuff.

The greatest gifts one can give to the outward world lie within. To reach them, you have to go there.

I'm not a technologist, so I can't say exactly how the outward bias of today's technologies might be changed. But the first step would be to adopt a different philosophical approach, one that acknowledges that in a busy, crowded world, less is more. That for many of life's most important and rewarding tasks, inwardness isn't just nice but essential. Perhaps on booting up, a digital device of the future might ask me how connected I want to be right now and offer various options, from alpha (less crowded, more focused) to omega (more crowded, less focused). If I chose alpha, it might then say "Choose one task" and not allow me to take on any others at the same time. A simplistic idea, perhaps, but then simplicity is what we need more of.

We're going to find it. Human beings are highly skilled at devising new ways to get away from the crowd. The recent past offers many examples. The Sony Walkman, the progenitor of today's digital music players, made the formerly outward experience of music inward and private as well as portable. Video-recording devices such as TiVo liberated the television experience from the constraints of time. Suddenly, it was no longer necessary to watch your favorite show when everyone else was or to endure those often annoying features of crowd life, commercials. The better such innovations serve the needs of the harried self, creating distance and space where there once was none, the more handsomely they're rewarded. Gutenberg's name is synonymous with the technology past, but as a business philosopher, he points straight to the future. In the long run, the smart money is on inwardness.

Chapter Eight

HAMLET'S BLACKBERRY

Shakespeare on the Beauty of Old Tools

"Don't worry," Hamlet's nifty device whispered, "you don't have to know everything. Just the few things that matter."

Several years ago, I bought myself some Moleskines, the simple, unassuming notebooks that have become popular in the last decade. By chance, my eye fell on the display at the front counter of my local bookstore. In particular, the "plain journal" model, which is roughly the size of a passport with tan cardboard covers reminiscent of paper grocery bags, was calling out to me. On impulse, I grabbed a three-pack and added it to my haul.

I'd been aware of the Moleskine phenomenon for years, ever since a globe-trotting journalist friend had shown me his. It was one of the classic larger models with pebbly black covers and an integral elastic band to keep it closed when not in use. The pages were filled with his lively handwriting and, scattered here and there, superb sketches of cats, people, and other subjects. The drawings were a surprise. I'd known this guy since we were teenagers and had never realized he had artistic leanings, let alone talent.

He brushed off my compliments and focused on the notebook

itself, which he said he couldn't live without. After the original French manufacturer had gone out of business in the late 1980s, wherever his travels took him he would comb the local shopping districts and bazaars for stray leftovers. That could go on for only so long, he knew, and he was preparing himself for the worst when, seemingly out of the blue, in 1998 an Italian firm brought back the Moleskine brand. To hear him tell the story, it was as if humanity had finally discovered the key to happiness and it would be smooth sailing from here on.

I opened the packet immediately, took one out, and held it. Though slender, it felt sturdy and substantial. The cardboard was a little fuzzy and faintly warm. I slipped it into my back right pocket, and I've had one there ever since. I pull it out at least a few times a day, to jot down ideas that come to me when I'm away from the computer I write on. Every other day or so, I look the notes over, cull the ones worth saving, and transcribe them into digital files. I also use the notebook for the occasional grocery list, driving directions, and other utilitarian tasks, but those scribbles go on the back pages, moving from the last page forward—a method Moleskine apparently had in mind, as the rear pages are perforated for easy tearing out. When the meaningful notes from the front meet the trivial ones from the back, it's time for a new notebook.

On the face of it, none of this makes much sense. In this seamlessly wired world, one can feel a bit loony toting around a stitched-together bundle of that dowdy, reportedly soon-to-be-obsolete tool, paper. There are so many more modern and efficient ways to record ideas and inspirations. For notes like mine that are headed for a hard drive anyway, it would be far more logical to go straight to digital, using a portable screen for the first step. Why bother writing them down by hand first? Another friend dictates thoughts he wants to remember into his smart phone, which automatically sends them as audio files to a transcription service. They're speedily e-mailed back to him as text, with, he reports, remarkably few typos. He loves this system and proselytizes for it. When I take my notebook out in his presence, he smiles the way one smiles at a passing Model T. *Look at that, isn't it sweet?*

Yet sometimes I think *I'm* on the cutting edge and he's stuck in the past. Just ten years ago, Moleskines were a rarity. Today I see them everywhere. When I'm using mine in public, someone nearby will often say, "You, too?" or "Aren't you just crazy about those things?" For bonding with strangers, it's almost as reliable as a baby or a dog.

I'm a true believer, but for a long time I didn't know why. What was it about this seemingly anachronistic tool that made me feel it was essential to my well-being? Why Moleskines, and why now? Their resurgence coincided exactly with the rise of digital connectedness, and my gut told me that the two must be related. But how? Was it just nostalgia, an effort to escape from the messiness of the present into the simplicity of an idealized past? Maybe paperphilia really isn't so different from the recessive pinings that motivate some people to own antique cars. I wanted to think there was more to it than that.

I found the answer in Renaissance England, of all places, with some help from that society's greatest creative mind, William Shakespeare. In the late sixteenth and early seventeenth centuries, London, where Shakespeare spent all his productive years and staged his plays, was a bustling, chaotic place. He and his contemporaries woke up each morning to a world as hectic and confounding in its own way as ours, and they found surprisingly inventive ways to cope. It was from one of their coping techniques that I came to see why, four hundred years later, it makes perfect sense that I find myself scribbling in vintage notebooks and feeling all the better for it.

One of the first plays Shakespeare ever wrote, *Henry VI, Part 2*, is about a mob of illiterate peasants that lays siege to London in an uprising against the wealthy and powerful. Having taken an important nobleman prisoner, the rebel leader accuses him of various crimes, among them that he's allowed printing presses to operate and spread knowledge of written language: "[T]hou hast caused printing to be used and . . . thou hast built a paper-mill. It will be proved to thy face that thou hast men about thee that usually talk of a noun and a verb and such abominable words as no Christian ear can endure to hear."

Far from seeing Gutenberg's invention as a liberating force, these rebels view it as a tool of oppression. There are good reasons for this. When Shakespeare wrote this play around 1590, he was a young man recently arrived in London from the provincial town of Stratford-upon-Avon. For its time, London was a sprawling metropolis, the third largest city in Europe with a population approaching 200,000. As Stephen Greenblatt writes in his Shakespeare biography, *Will in the World: How Shakespeare Became Shakespeare*, he must have been stunned by "the London crowd—the unprecedented concentration of bodies jostling through the narrow streets, crossing and recrossing the great bridge, pressing into taverns and churches and theaters . . . their noise, the smell of their breath, their rowdiness and potential for violence."

There was a palpable undercurrent of danger, too, and not just from street crime, the ever-present threat of plague, and other perils endemic to urban life in this period. This was also a politically unstable time and place. Ever since King Henry VIII's break with the Catholic Church earlier in the century, England had been a pressure cooker of political tension, with power swinging back and forth between Protestants and Catholics, depending on who was on the throne. By the time Shakespeare turned up in London, the Protestant Queen

Elizabeth I had been ruling for many years, her regime always on the watch for Catholic dissidents and spies. Heads were being chopped off all the time, in some cases only to reappear later in a ghastly lineup on the gates of London Bridge. In *The Prince and the Pauper*, his novel of sixteenth-century London, Mark Twain describes this practice, noting that "the livid and decaying heads of renowned men impaled upon iron spikes" were put on display at the bridge as "object lessons" for passersby. Be careful, they said, or this could be you.

In this treacherous atmosphere, the idea of a play about a violent rabble trying to overthrow the powers that be makes more sense. But it still doesn't explain the rebels' specific hostility to Gutenberg's invention, a plot element that could not have come from the historical revolt on which the play is based, which took place in the fourteenth century, before the press even existed. It was Shakespeare who worked this detail into the story. Like Gutenberg, he left behind few clues about his life and the thinking that drove his work. However, we do know that he lived in a rapidly changing world, and one of the key forces driving the change was technology. The printing press was transforming society in countless ways, and, as with digital technology, some of the changes were a source of anxiety and tension.

The printing press had dramatically increased book production. There were an estimated 8 million printed copies of books in Europe by 1500 and far more by Shakespeare's time. In most respects this was a tremendously positive development. As more people learned to read and gained access to books, the opportunities for individual growth and advancement multiplied, which in the long run could only be good for the world. There's no better example of this than Shakespeare himself, who was born into a community where few could read and one day would be called "the poet of the human race."

But it can take a long time for a society to adjust to a powerful new technology and figure out its best uses. And for all the salutary effects of print, it presented challenges. Though today we know the printing press played a crucial role in the rise of individualism and democracy, like any powerful medium it was sometimes used as a tool of social and political control. While books were more plentiful and accessible than ever, they remained expensive, and literacy was far from universal. As the ability to read took on greater importance, the divide between those who could read and those who couldn't was felt often in everyday life, sometimes in disturbing ways. For instance, English law at the time made a distinction between literate people accused of committing felonies and illiterate ones. In certain cases, accused criminals who could read were tried in ecclesiastical courts, which did not impose the death penalty. The illiterate, meanwhile, went before government courts, where death was a frequent punishment. In effect, people were hanged as a direct result of the fact that they couldn't read.

In this context, as Greenblatt observes, one can see how a playwright intimately familiar with both worlds, literate and not, could imagine a gang of unlettered ruffians wanting to destroy the presses. Print represented power, and the nonlettered had cause for resentment.

More broadly, this technology simply increased how much there was to know and process. Some 1,500 years after Seneca complained about the burden of all those books, the hunted-mind syndrome was back in a more intense way. Once again, there had been a colossal expansion in the sheer amount of available information, without a matching increase in the capacity of the human mind to absorb it. Beyond books, Europe was awash in pamphlets, advertising placards, commercial and public documents—bureaucracy mushroomed wildly in this period—and many other types of printed matter. The first newspapers were about to be launched. There was a lot to handle, and everyone, even the illiterate, felt its effects. Then as now, it was unsettling just knowing it was all out there. As modern-day scholar Ann Blair has shown, those who lived through the Renaissance experienced something very much akin to the overload we feel today.

What did they do about it? The answer lies a decade further along in Shakespeare's career, in the most familiar and resonant of his plays, *The Tragedy of Hamlet, Prince of Denmark*. There's a moment in *Hamlet* that speaks to our technological dilemma and helps explains the curious persistence of paper notebooks in a digital world. In Act I, Hamlet meets the ghost of his dead father, whom everyone at this point believes was killed by a serpent's bite. The ghost has a news flash: he wasn't done in by a snake but poisoned by his brother, Hamlet's uncle Claudius, who is now king. The ghost beseeches Hamlet to avenge this "murder most foul," and bids him a spooky farewell, "Adieu, adieu, Hamlet. Remember me."

Hamlet's reaction to all this is a bit surprising. Rather than focus on the ghost's staggering message, he reflects on his own state of mind and in particular his memory:

Remember thee?
Ay, thou poor ghost, while memory holds a seat
In this distracted globe.

The distracted "globe" he's talking about is his head—the actor playing Hamlet would have grasped his own while reciting the lines. Yes, he's saying, of course I'll remember you, because somewhere in this chaotic, unruly brain, I do still have a memory. Shakespeare also seems to be punning off two other meanings of "globe." On one level, he's suggesting that the whole world is distracted, and on another, that the audience watching the play in the Globe Theatre might be having some mindfulness issues. Attention deficit disorder was apparently raging long before we gave it that inelegant name.

Hamlet goes on:

> Yea, from the table of my memory
> I'll wipe away all trivial fond records,
> All saws of books, all forms, all pressures past,
> That youth and observation copied there,
> And thy commandment all alone shall live
> Within the book and volume of my brain
> Unmixed with baser matter.

What he's basically saying is "I'm going to clean out all the mental clutter that makes me so distracted, so the only thing taking up space in my head will be you, Mr. Ghost, and this hideous crime." Notice that twice in the above lines Shakespeare has Hamlet mention books, but with a different meaning in each case. In the third line, books (and the "saws," or clichés, that that they contain) are part of the detritus and "pressures" he needs to remove from his mind so he can think clearly. Then, just a few lines later, he likens the brain itself to a book, a very appealing one devoted to a single important subject and absolutely free of worthless trivia ("baser matter"). In effect, he's saying he's determined to cure his own mental overload by throwing out the equivalent of a lot of books in order to make room for the one book that really matters, his mind.

Shakespeare had an intense interest in books, as one would expect of a man whose life was shaped by them, and they figure often in his works. There's another moment in *Hamlet* where the stage directions call for the prince to enter reading a book. From the above passage, it's clear that the playwright had a nuanced understanding of the wide range of effects books can have on a person. A book could be a huge obstacle to clear thinking or it could be a tremendous help, depending on how one used it.

But why, amid all this talk of books, does Shakespeare also throw in a table? In the first line of the same passage above, Hamlet compares his memory to a table he's going to wipe off. When modern audiences hear the word "table," we think of the four-legged kind that figures prominently in our kitchens and dining rooms. And since we do wipe off those tables, the image initially makes sense. In fact, when Shakespeare used the word "table" in these lines he wasn't thinking of a piece of furniture. He was thinking of a piece of technology. Several lines down, the word returns:

My tables—meet it is I set it down
That one may smile and smile and be a villain.
At least I'm sure it may be so in Denmark.
    [*He writes.*]

Here Hamlet is marveling that King Claudius can walk around with a smile on his face even though he's a cold-blooded assassin. This strikes him as a thought worth remembering, one that it would be wise (or "meet") to record ("set down"). If you've seen the play performed, you may have watched the actor playing Hamlet scribble a note at this point. However, you likely didn't realize that the object he pulled out for this purpose was the one he was referring to when he said "My tables," as well as in the earlier passage about wiping away the mental clutter.

What *are* these tables, anyway?

They were an innovative gadget that first appeared in Europe in the late fifteenth century. Also known as writing tables or table books, they were pocket-sized almanacs or calendars that came with blank pages made of specially coated paper or parchment. Those pages could be written on with a metal stylus and later erased with a sponge, so they were reusable. Tables were a new, improved version of a technology—wax tablets—that had been around for centuries. Instead of wax, their surfaces were made of a plasterlike material that made them much more durable and useful. They became enormously popular in Shakespeare's lifetime as a solution to the relentless busyness of life. A harried Londoner or Parisian would carry one everywhere, jotting down useful information and quick thoughts, perhaps checking off items on a to-do list.

We don't know that Shakespeare owned a table himself, but since he took the trouble to insert one into *Hamlet* and they were very popular among people in his world, it's not unreasonable to imagine he did. It would have been useful to a man who was not only constantly writing plays (and collecting words and phrases to use in them) but also acting (he played the ghost in *Hamlet*), running a business of which he was part owner (the Globe), and investing in real estate on the side, all while trying to stay in touch with distant friends and family—his wife and children remained in Stratford, never coming to live with him in London. Never mind the sonnets, the alleged love affairs, and who knows what else. He had a lot on his plate, and for anyone who did, this technology was a godsend. It was a portable, convenient way to manage the endless details of an active life, the period equivalent of our BlackBerrys and iPhones.

According to a scholarly article published in the *Shakespeare Quarterly* in 2004, the many uses of tables included:

collecting pieces of poetry, noteworthy epigrams, and new words; recording sermons, legal proceedings, or parliamentary debates; jotting down conversations,

recipes, cures, and jokes; keeping financial records; re-
calling addresses and meetings; and collecting notes on
foreign customs while traveling.

Users spoke effusively about their tables and swore they
couldn't get by without them. Michel de Montaigne, the great
French essayist who was roughly contemporary with Shakespeare, said it was impossible for him to make his way through
a complicated discourse with another person "except I have my
writing tables about me" for jotting notes. "Yes, sure I never
go without Tables," says a character in an early-seventeenth-
century play by Edward Sharpham. Tables migrated to the
New World and caught on there, too. Thomas Jefferson owned
one, and they remained popular into the nineteenth century.

Given that it played a central role in people's lives for
hundreds of years and helped some of history's most brilliant minds organize their time and thoughts, it's remarkable
that this device has been almost completely forgotten. In fact,
Hamlet's trusty handheld has a few messages for us.

ONE OF THE most widely held assumptions of modern culture
is that when a new technology comes along, it automatically
renders obsolete the older ones that performed roughly the
same function. The classic case is the buggy whip. When society switched from carriages to automobiles in the early twentieth century, there was no longer any need for buggy whips,
and they effectively disappeared. However, it doesn't always
work this way. Older technologies often survive the introduction of newer ones, when they perform useful tasks in ways
that the new devices can't match.

The best example is the hinged door. Watch a science fic-
tion movie some time and pay close attention. You'll notice

that the houses, office buildings, and spaceships of "the future" almost always have sliding doors. Since the 1920s, filmmakers have assumed that in the future there would be no hinged doors whatsoever. Why? Because hinges are old-fashioned and cumbersome. The doors that swing on them take up a lot of space. There's no good reason we should continue to use this antiquated, literally creaking technology when sliding doors make so much more sense. They're so sleek and logical and, well, futuristic. Thus, in the popular imagination hinges are always on the verge of extinction.

Yet, as you've undoubtedly noticed, hinged doors are still very much with us. Why? Because though sliding doors are aesthetically appealing, when you come down to it, they do only one thing, slide in and out, which is kind of boring. Hinged doors are more interesting precisely because of the way they occupy and move through space. You can burst through one and surprise somebody. You can slam a hinged door loudly to vent your anger or close it very quietly out of concern for a sleeping child. A hinged door is an expressive tool. It works with our bodies in ways that sliding doors don't.

"Time has given the hinge a rich social complexity that those who foresee its imminent demise fail to appreciate," writes Paul Duguid, a scholar and author who has used the hinge to demonstrate that new technologies don't always vanquish or supersede old ones.[*]

In some instances, an older technology will survive not just by doing what it's always done well but also by taking on a brand-new role. When television arrived in the 1950s, many expected radio to disappear. Why would you want an old box that produces only sound, when you could have a new one with sound *and* video? In fact, television did replace radio as the

---

[*] *There are societies, notably Japan, where sliding doors have long been popular.*

dominant medium for news and entertainment and as a gathering place in the home. But radio found new roles to play. In the automobile, for example, where drivers were not in a position to watch video, radio was a natural choice. Today, in our information-jammed world, many of us enjoy radio precisely because it produces only sound—no text, images, or video—and can relieve media overload.

What does this have to do with Shakespeare? I likened Hamlet's erasable table to the smart phones we carry around today because, like the latter, it was a new gadget that helped people better manage their busy lives. However, it was a new gadget built on two very old technologies. I've already mentioned one, the older wax-based device. The other, much older technology was handwriting. Remember, this was a time when handwritten communication was, in certain crucial ways, on the decline. After centuries of handwritten texts, Gutenberg had come up with a much more efficient technology. People had immediately recognized the value of his invention, and printing had taken off. According to the sliding-door school of thought, then, by Shakespeare's time handwriting should have been relegated to a much smaller role in society and everyday life.

In fact, the opposite happened. Though hand-produced manuscripts did go into a long, slow decline, beyond the small world of professional scribes the arrival of print set off a tremendous popular expansion in handwriting. Even as the revolutionary new Gutenberg technology was taking hold—and in some ways *because* it was taking hold—the older one gained new life. There were a couple of reasons for this. First, as printed matter become widely available, the very idea of engaging in written expression suddenly became thinkable to more people. Previously, putting one's own ideas into words on a page had been the province of the rich and powerful. With printed texts flying around everywhere, this rarefied activity looked less ex-clusive and intimidating and more appealing. Regular people wanted and often needed to participate in this new conversation. Since most didn't have access to a press, handwriting was the best way to join in. Many who couldn't read or write were suddenly motivated to learn.

"The advent of printing was a radical incitement to write, rather than a signal of the demise of handwritten texts," write Peter Stallybrass, Michael Mendle, and Heather Wolfe, authors of groundbreaking scholarship on this phenomenon. As a result, all sorts of important new technologies for writing by hand appeared after the printing press, including graphite pencils and fountain pens. Print simply made more people want to write.

The second reason handwriting became so popular was that it turned out to be a very useful way to navigate the whirlwind of information loosed by print—to live in a crazy world without going crazy oneself. New shorthand methods were invented for taking down words more efficiently. The script style called "round hand," the forerunner of the cursive writing we use today, was created during this period, for the same reason.

But the most compelling example of handwriting's ability to lighten the burdens of the post-Gutenberg mind was the gadget that Shakespeare gave Hamlet. Here was a fantastic antidote to the new busyness, a portable, easy-to-operate device that allowed the user to impose order on the clamorous world around him. Hamlet wasn't the only one with a tumultuous, tricky-to-navigate life. Imagine Shakespeare back at home after a busy day at the Globe and perhaps some helter-skelter errands around town. At some point in the evening, maybe just before bedtime, he takes out his tables and reviews everything he's written there since morning. He pulls the stylus out of its handy hidden groove in the binding and circles the jots he wants to keep, while *X*-ing out those that can be tossed. He transcribes each of the keepers to the appropriate hard-copy volume, which might be a diary, a commonplace book for saved quotes and scraps of language, or a financial accounts book. When he's done, he takes a small sponge (or a wet fingertip) and erases the surface of the pages so they're ready for the next day. And no charger to plug in!

The easy erasability of tables was central to their success. In an epoch when so many words were being committed permanently to the printed page—more than any one mind could handle—this gadget moved in exactly the opposite direction. At the owner's command, it made words go away, vanish, cease weighing on the soul. "Don't worry," Hamlet's nifty device whispered, "you don't have to know *everything*. Just the few things that matter."

In effect, it was a pushback against all the people and information that were closing in, or often seemed to be. With one of these in your pocket, you were in the driver's seat. You could be selective about what you brought home with you—both literally home to your dwelling and figuratively home to your mind. It was a surrogate for the mind, a visible, tangible representation of what was going on inside your own "globe," and a way of improving its performance. This is essentially what Hamlet vows to do when he likens his own mind to a table that he's erasing, so he can focus on Topic A. He is cleaning up the internal mess, starting fresh.

I had a chance to see some genuine tables from the period at the Folger Shakespeare Library in Washington, D.C. One of them, made in London within a few years of the debut of *Hamlet*, had the user instructions still intact ("To make cleane your Tables, when they are written on, Take a lyttle peece of Spunge . . ."). I could almost visualize it for sale at the front

counter of a modern bookstore. Like these four-hundred-
year-old tables, my Moleskines rely on old tools, handwriting
and paper, that in a world of clicking keys and glowing screens
are widely assumed to be nearly obsolete. Yet both are central
to why this humble tool gives me a sense of mental order and
control.

Unlike my screens, which thrust words, images, and
sounds at me all day and night, my paper notebooks project
no information at all. The pages are blank. They invite me
to fill them with information, and when I do, it's information
of my own choosing that I write with my own hand. Crossing my front yard one morning, for instance, I remembered
an obscure historical fact about Madagascar that I'd heard the
day before and realized might be useful in a writing project
I've got on the back burner. Out came the notebook, in went
Madagascar. Having survived the winnowing processes of my
consciousness, it had earned a spot on the page, and just the
act of writing it down raised its profile in my thoughts. When
you're used to clicking keys all day, shaping letters one by one
feels exotically earthy, memorable just by contrast.

Digital screens are tools of selectivity, too, but using them
is more reactive, a matter of fending off and filtering. Because
a paper notebook isn't connected to the grid, there's no such
defensiveness. The selectivity is autonomous and entirely self-
directed. I'm the search engine, the algorithm, and the filter.
(Which is not to say it feels like hard work. Sometimes I just
doodle.) Like tables, my notebooks are a pushback against the
psychic burden of a newly dominant technology.

However, there's an important difference. In the sixteenth
century, when information was *physically* piling up everywhere,
it was the ability to erase some of it that afforded a sense of
empowerment and control. In contrast, the digital information that weighs on us today exists in a *nonphysical* medium,

and this is part of the problem. We know it's out there, and we have words to represent and quantify it. An exabyte, for instance, is a million million megabytes. But that doesn't mean much to me. Where *is* all that data, exactly? It's everywhere and nowhere at the same time. We're physical creatures who perceive and know the world through our bodies, yet we now spend much of our time in a universe of disembodied information. It doesn't live here with us, we just peer at it through a two-dimensional screen. At a very deep level of the consciousness, this is arduous and draining.

It's a complex problem that my notebook addresses with utter simplicity. My first Moleskine purchase was driven largely by its tactile appeal. I wanted to feel it in my hands and flip its creamy pages with my fingers. I wanted to interact with it in ways in which I never get to interact with my screens. But the draw isn't only sensuous; it's also about physical *presence*.

In conventional thinking about technology today, the fact that paper is a three-dimensional medium—that it's made of atoms rather than bits and therefore takes up space—is considered its great weakness. Like you and me, it has a body and is stuck here in the physical world. My notebook can't fly from here to China in seconds the way digital data can. However, just as the strength of digital devices (their ability to bring the crowd closer) is also their weakness, the weakness of paper can be a strength.

Among researchers who study how humans interact with technology, there's a theory known as embodied interaction, which says that three-dimensional tools are easier on the mind in certain important ways. This makes intuitive sense. Think of a screen with a dozen different documents open, all layered on top of one another, and what a pain it is try to organize and keep track of them all at once, using just your clicker and keyboard. Sometimes you want to reach in there and grab them, but you can't. Reading and writing on screen, we expend a great deal of mental energy just navigating. Paper's tangibility allows the hands and fingers to take over much of the navigational burden, freeing the brain to think. Because a notebook has a body, it works more naturally with our bodies. At a time when movies and other screen experiences are striving for 3-D effects, paper is in one sense ahead of the curve.

In this high-speed era, another plus is the simple fact that my notebook *isn't* connected to the electronic grid. It slows down information, gives it a resting place. The process of writing and thinking on screen has a wonderful lightness, a sense of constant changeability and evanescence. But sometimes you need to touch down. As the pamphlet that comes tucked into each Moleskine says, it's a way "to capture reality on the move." I can pull ideas not only out of my mind but out of the ethereal digital dimension and give them material presence and stability. *Yes, you exist, you are worthy of this world*. It doesn't matter that the best of my notes will ultimately reside on my hard drive. The point is that before any of that happens, while the ideas are still cooking, I spend time with a tool that brings out the best in my mind. It may be an old tool, but, like a hinged door, it can do things that the new gadgets can't.

What Hamlet's tables and my notebooks share is this: each is an effective way of bringing an unruly, confusing world of stimuli and information under control. In Shakespeare's time, it was the bedlam of the crowded city and the pressures of the emerging print culture. Today, it's the bottomless inbox, the ringing cell phones, and just the weight of all that weightless digital stuff. Either way, the issue is overwhelming connectedness and the fundamental human need to work a little disconnectedness back into the equation.

Over and over in history, new technologies arrive that play to our natural maximalist tendencies. At the same time, quietly but persistently, there's a need to find balance. The best solutions serve as a kind of bridge to the tech future, one that ensures that we'll arrive with our sanity intact.

Stephen Greenblatt writes that one of the great achievements of *Hamlet* is its "intense representation of inwardness." Shakespeare found a new way to capture real thought, what actually happens in the mind of an individual as he wrestles with a problem. Hamlet's inwardness is the essence of the play's power, and when he takes out his tables, inward is where he's headed. Having received a jolt from the outward world (of which the ghost, being otherwordly, is a perfect representative), it's where he needs to go. He says as much when he speaks about bringing order to "this distracted globe" by getting rid of all those other books, leaving only "the book . . . of my brain." When he's done setting down his thoughts about his terrible uncle, he puts his gadget away, saying, with new resolve, "So uncle, there you are. Now to my word." In other words, back to the task at hand, his promise to seek vengeance.

Hamlet winds up having some trouble sticking to that task, of course, and things end badly for him. Who knows, perhaps if he'd used his tables a little more regularly—they never return for the rest of the play—it might have worked out differently.

Chapter Nine

INVENTING YOUR LIFE

Ben Franklin on Positive Rituals

"All new tools require some practice before we can become expert in the use of them."

One idea for improving digital life that seemed promising from the start was no-e-mail Fridays. Businesses and other organizations had experimented with various versions of the concept before it drew wide attention several years ago, when numerous studies and media reports cited it as a possible answer to the problem of distracted, inefficient workers. Something had to be done, with the losses in productivity from information overload estimated to be in the hundreds of billions of dollars.

The notion was appealing because it was so simple: one day a week, everyone in an organization would disconnect from the inbox. This would lighten the mental burden, restore focus, and encourage face-to-face interaction. It echoed the casual Fridays that caught on in the workplace in the late nineties and changed long-standing attitudes toward one aspect of office behavior, how people dressed. If it worked for clothing, why not for the screen? After a few months of ritualistically avoiding the inbox one day a week, it would become second nature. Workers would be able to ease off any time they needed a break, not just on Fridays. Problem solved.

Yet it hasn't happened. Though there have been scattered reports of organizations successfully implementing such regimens, they have not been widely adopted or embraced. To the contrary, stories abound of workers either openly resisting the rules or covertly cheating. The hiatus is aimed at helping them, but they don't seem to view it that way.

"Withdraw it even for a day, and some employees fight back like recovering smokers in a nicotine fit," reported the *Wall Street Journal*. At one California technology company, a new e-mail moratorium had been in effect for less than fifteen minutes when a worker who normally sent hundreds of e-mails a day couldn't take it any longer and fell off the wagon. "It's kind of like speeding," he told the newspaper. "You know there's a law that says you're not supposed to do it, but when you're in the heat of combat, you aim and fire."

The image speaks volumes, reflecting how digital technology has turned the workplace into a war zone for the mind. What ban-breaking workers effectively are saying is that they have no choice but to stay hunkered down in their foxholes. After all, just because management has established a no-e-mail day, that doesn't mean e-mail stops arriving. So if you're away from the screen for a whole day, you fall behind in your work, which nobody likes to do. And for many tasks e-mail really is the easiest, most efficient means of communication.

More broadly, this is how workers live now, at the office and everywhere else—always connected. And once in screen mode, it's very hard to get out. It's this digital inertia that's responsible, at least in part, for the perpetual haze—and the poor work habits—employers are battling.

We've all been on the receiving end of this problem,

whether it's a customer service phone rep who has to be asked the same question three times or a retail store salesperson who abandons you for her handheld. You think: *This company has to be crazy, to employ someone this unfocused.* But this is the new normal. Distracted students sit in screen-lined classrooms half listening to distracted teachers. Drivers race through red lights and cross median strips, killing themselves or others, for the sake of getting out another text message. We take the fog of the war zone home with us and, by keeping one eye on the screen, ensure that it never lifts.

There's a compulsive inevitability to the cycle and, in some pessimistic quarters, a fear that it can't be broken. I've heard middle-aged people grumble that digital natives (those roughly thirty and under, who have grown up with screens) are effectively a new species of human being, innately incapable of holding a sustained conversation or thought. *Homo distractus.* This is the future, they moan, get used to it, as they sneak a glance at their own mobile.

Defeatism will only make these problems fester, when they don't have to. The struggle of employers to impose some sanity on the e-mail burden is a reminder that, in the broad scheme of things, *all* digital technologies are still very new and we're just in the early stages of figuring them out.

We take for granted how easy it is to use older devices, though many of them had long adjustment periods. When the telephone arrived in the late nineteenth century, it was widely viewed as a passive, one-way listening contraption. In Europe, phone service was initially marketed to the public as a way to enjoy opera and other live performances without attending in person. Of course, phones ultimately became a person-to-person device, but even then, how we use them has evolved over time. For much of the twentieth century, when the phone rang it was customary to drop whatever you were doing and answer it—people were slaves to the bell. In old movies, you know a businessman or some other high-powered character is way too busy when he's shown speaking on two or more phones at once. When answering machines and voice mail came along several decades ago, it was a real advance in everyday life. And we're *still* learning to live with phones. Lately the lapses in common sense and manners that characterized the early mobile phone years—people taking calls during Broadway shows—seem to be receding, a sign that we're sorting out at least some questions.

But the learning curve never ends. With every new device, there are three categories of issues. First, the purely *functional*: What can this device do for us? What are its best uses? Second, the *behavioral*: Are there old behaviors I need to change or new ones I need to acquire in response to this? These are all exterior questions, but beneath the surface there's a third category that's often ignored, especially early on: the inner *human* dimension of technology: How is this device affecting me and my experience? Is it altering how I think and feel? Is it changing the rhythms of my day? Does life seem to be moving more quickly (or slowly) as a result of this gadget? Is it affecting my work? My home life? If so, are the effects good or bad?

These human issues are the ones that ultimately matter most, and it's when problems arise in this area that we really begin to question the ways we're using a technology and look for new approaches. True, big businesses began to worry about digital overload only when they saw it was costing them money. But the reason it showed up in the bottom line was an entirely human one: the *minds* of workers were off kilter. To ignore the interior is to set yourself up for trouble.

Since no-e-mail rituals are aimed squarely at the mind, why haven't they worked? In fact, regimens of this kind *are* a solid idea, based on the age-old insight that rituals form good habits, which in turn are the foundation of a fruitful, happy life. Aristotle said it more than two thousand years ago: "We are what we repeatedly do. Excellence, then, is not an act but a habit." But simply engaging in a ritual—performing a given action at a certain time and/or in a particular way—is not enough. The transformative power of rituals is rooted in what they mean *to the person performing them*. In order to change a deep-seated, habitual behavior through ritual, the individual must believe that he or she needs to change. It's about not just *how* but *why*. Inner change depends on inner conviction.

Nobody sheds light on this question better than Benjamin Franklin, who embodied all the qualities that no-e-mail Fridays and other well-intentioned digital diets seek to encourage. Franklin was a model of mental clarity and productivity, despite the fact that he lived an intensely busy life and was constantly juggling responsibilities and projects. And he was a great champion of rituals, which he used to fight his own worst tendencies and bring his overextended life under control. He attributed his many achievements in business, government, science, and other fields, and the contentment he enjoyed along the way, to one particular ritual that he developed early in life.

Today's efforts to wean workers from the screen often assume that it's possible to eliminate negative traits and tendencies just by imposing rituals from above. *If everyone here at Acme Widget stops e-mailing on Fridays, we'll all become less addicted to our inboxes.* Franklin took a different approach: he identified not just the negative character traits he wanted to change but the positive, inward reasons why he wanted to change them. Only then did he begin modifying his behavior toward those ends, through a self-designed ritual. The conviction came first, and it made all the difference.

. . . .

IN THE SUMMER of 1726, the twenty-year-old Franklin sailed home to Philadelphia from London, where he'd spent the last few years working in the printing business. Gutenberg's technology was now nearly three hundred years old, and the confusion and class tension it had engendered in Shakespeare's time were largely gone. Literacy was far from universal, but more widespread. Though Franklin came from a modest background—his father was a tallow chandler, a maker of candles and soap—he had grown up reading hungrily and gone into Gutenberg's trade himself.

Franklin's era was also different from those discussed in previous chapters in that it was not a time when a new connective technology was shaking up society and the mind. In the eighteenth century, people and their ideas moved through time and space largely by well-established means: on foot, on horseback, in carriages and ships; via messengers and mail services; and through print media, including newspapers, pamphlets, and books. The leading scientists of the time, including Franklin himself, would lay the groundwork for spectacular technologies to come, but those were off in the future.

If technology per se wasn't bringing the crowd closer, it was a very busy time in other ways. Franklin's life coincided with the Enlightenment, an age of great cultural and intellectual ferment that stressed the power of human reason to discover truths about existence, as opposed to relying on religion and the received wisdom of the past. Great thinkers such as Isaac Newton, Voltaire, Thomas Paine, and Adam Smith were exploring bold new ideas in science, philosophy, politics, economics, and other fields. Among the results of this new thinking would be two revolutions, the American and the French. As a young man, Franklin eagerly waded into the outward sphere where this vibrant conversation was unfolding and stayed there, living the rest of his life essentially in public. Intensely gregarious by nature, he was drawn to the crowd and loathed being away from it and, in this way, very much a man of his age. "One of the fundamental sentiments of the Enlightenment was that there is a sociable affinity . . . among fellow humans," writes his biographer Walter Isaacson, "and Franklin was an exemplar of this outlook."

This is what makes his story valuable today. Even without a nudge from new technology, he had a persistent yen to connect that was very much like the one that drives us back to the screen and the busyness it delivers. He was constantly out and about doing the eighteenth-century equivalent of social networking. One of the pseudonyms Franklin used in his early years as a newspaper writer, the "Busy-Body," described him perfectly.

On the long voyage from London, however, he had time for some philosophical soul-searching, and he came away with two insights that would resound throughout his life. To pass the hours, he and his fellow passengers played cards. A few weeks into the trip, it was discovered that one of them was cheating. As punishment, the group demanded that he pay a fine of two bottles of brandy. When he refused, they "excommunicated" him, as Franklin later wrote, "refusing either to play, eat, drink, or converse with him." The miscreant quickly paid up, for reasons Franklin well understood and explained in his shipboard journal:

> Man is a sociable being, and it is . . . one of the worst punishments to be excluded from society. I have read abundance of fine things on the subject of solitude, and I know 'tis a common boast in the mouths of those that affect to be thought wise, that they are never less alone

than when alone. I acknowledge solitude an agreeable
refreshment to a busy mind; but were these thinking
people obliged to be always alone, I am apt to think
they would quickly find their very being insupportable
to them.

"Give me the crowd any day," in other words. Franklin
was justifying his own craving for nonstop connectedness. On
the same voyage, however, he had a second important insight
that cut in a different direction. He realized that his life, as
he'd been leading it so far, was terribly disorganized and out
of balance. He wasn't good at managing money and relationships, and his career wasn't headed where he wanted it to go.
And he knew the problem: he was racing around in too many
directions. "I have never fixed a regular design as to life," he
wrote, and rather than a coherent whole, "it has been a confused variety of different scenes." Spending so much time at
the crowded, omega end of the continuum of connectedness
was not bringing him happiness or success. Like many of us,
he was a victim of his own busyness.

In a era when bold new ideas about human freedom were
in the air, Franklin saw that to be truly free, one must con-
quer not just outward oppressors but inward ones—the habits
that keep us from achieving all that we might. It was time to
focus his scattered self, and he came up with a ritual to do just
that. He realized that the essential problem was that he had
not learned to control his impulses. His crowded outward life
paraded many temptations before him, and he gave in to them
more than was healthy—his sex drive, for example, often got
him into trouble. Others had religion to guide their behavior,
but Franklin was a skeptic who belonged to no church. He
was, however, a great fan of philosophy, and to address his dilemma, that's where he turned.

He wrote two fictional dialogues, modeled on Plato's, in which a fellow named Horatio who can't resist his impulses is arguing with his friend Philocles, who is guided by reason. Horatio says one should always obey one's urges, because they're natural. To deny ourselves what we instinctively crave is absurd and wrong.

Philocles replies that it's Horatio who has it all wrong. Self-denial is a route to *greater* pleasure than you can ever obtain by just obeying your desires.

This sounds fabulous to the pleasure-loving Horatio, who asks his friend to explain how exactly it works.

Philocles says it's just a matter of refusing do something that you know is "inconsistent with your health, fortunes, or circumstances in the world; or in other words, because 'twould cost you more than 'twas worth. You would lose by it, as a man of pleasure."

Far better, he suggests, to control and manage your impulses by practicing what he calls "philosophical self-denial"—resisting certain urges because you know you'll gain more in the big picture by doing so. The operative word here is "philosophical." Franklin was saying that in order for self-denial to work, you have to reason it out first in your own mind. You have to see that there is more to be gained by resisting the impulse than giving in. Once you truly believe this, it's all downhill. What previously seemed a dreary, priggish way to live—denying oneself pleasure—suddenly becomes positive and even hedonistic.

This pragmatic approach appealed to Franklin, who was nothing if not practical. He published the dialogues in the newspaper he'd recently founded, then sat down and devised an ambitious self-improvement plan for himself. Rather than just swearing off his bad habits, he would practice philosophical self-denial. He looked inside himself and figured out what were the good habits that, if acquired, would cancel out the bad ones *and* make his life a lot nicer. He wrote down thirteen desirable virtues along with behavioral guidelines for attaining each one:

1. **Temperance** Eat not to Dullness. Drink not to Elevation.
2. **Silence** Speak not but what may benefit others or your self. Avoid trifling Conversation.
3. **Order** Let all your Things have their Places. Let each Part of your Business have its Time.
4. **Resolution** Resolve to perform what you ought. Perform without fail what you resolve.
5. **Frugality** Make no Expence but to do good to others or yourself: i.e. Waste nothing.
6. **Industry** Lose no Time. Be always employ'd in something useful. Cut off all unnecessary Actions.
7. **Sincerity** Use no hurtful Deceit. Think innocently and justly; and, if you speak, speak accordingly.
8. **Justice** Wrong none, by doing Injuries or omitting the Benefits that are your Duty.
9. **Moderation** Avoid Extremes. Forbear resenting Injuries so much as you think they deserve.
10. **Cleanliness** Tolerate no Uncleanness in Body, Clothes or Habitation.
11. **Tranquility** Be not disturbed at Trifles, or at Accidents common or unavoidable.
12. **Chastity** Rarely use Venery but for Health or Offspring; Never to Dullness, Weakness, or the Injury of your own or another's Peace or Reputation.
13. **Humility** Imitate Jesus and Socrates.

He needed a method for pursuing these goals and tracking his progress. So, in an ivory version of Hamlet's erasable tables that he carried everywhere, he drew up a series of elaborate charts, one for each virtue. Every day he marked down how he'd done. If it was a bad day for, say, Frugality, he'd make a black mark next to that goal.

He called this ritual "the bold and arduous Project of arriving at moral Perfection," and, as the label suggests, it was ridiculously ambitious. A saint would have trouble sticking to Franklin's program, though he did give himself wiggle room on venery—sex—with a broad exception for "health." He eventually saw that he'd overdone it and loosened up his standards, and after some years he stopped updating the charts altogether. But he carried them around with him for the rest of his life, a tangible reminder of what he was still aiming for, if not always expecting to achieve.

Toward the end of his life, looking back in his autobiography, he said that the ritual had made him who he was. "This little Artifice," as he called it, had instilled in him the habits that were responsible for everything he valued most in life, including his health, his financial success, and all his remarkable achievements. Franklin didn't have just one brilliant career, he had half a dozen. He was a skillful businessman, a trailblazing journalist and writer, a prolific scientist and inventor, an influential public official, and a political thinker whose ideas contributed significantly to the birth of modern democracy. To have done any one of these things would have been impressive. That he did all of them, and enjoyed himself along the way, is miraculous. Though absurdly busy, he calmly marshaled his time, talents, and energies to serve his goals. "Franklin's powers were from first to last in a flexible equilibrium. . . . He moved through his world in a humorous mastery of it," writes his biographer Carl Van Doren. And Franklin chalked it all up to his ritual, urging others to "follow the Example & reap the Benefit."

In the centuries since, many have derided the project as puritanical, sanctimonious, and self-congratulatory. "He made himself a list of virtues, which he trotted inside like a grey nag in a paddock," wrote the novelist D. H. Lawrence. If the idea had come from a straitlaced prude, it would be unbearable. But because Franklin *wasn't* a Puritan—he loved his pleasures—and had such a sense of humor about himself, his account of it is delightful to read today. It's also instructive. At a moment when so many are struggling to rein in one particular impulse, the question is: why don't *our* rituals have the same kind of success?

Franklin understood human nature, and he recognized that in order for a ritual to succeed, people have to believe in it. But belief can't be imposed by the world at large or higher-ups in management. It has to come from within. That's what "philosophical self-denial" is all about. To change a given habit, people must believe that by changing they'll gain more than they would by sticking to their old ways. Franklin applied this principle directly to himself. The virtues on his list are positive goals that he had concluded, through introspection, would bring him greater happiness than he would enjoy without them.

Thus, rather than calling his first goal "Stop Drinking So Much," a tut-tutting negative that would only drive home what he was losing out on, he called it Temperance. Why? Because he *liked* to drink and he needed an upbeat objective that he could embrace in the belief he was going to come out ahead. Temperance isn't quitting cold turkey; it's just moderation, a worthy and pleasant aim. The whole list works this way, emphasizing the positive—the instructions on what *not* to do are subordinate—to reinforce his own engagement. He would mend his chatterbox ways not merely by talking less but by actively seeking Silence, an appealing objective. Instead of just ignoring the trivia or "trifles" that can be so distracting, he'd be pursuing Tranquility, and who doesn't want that? In effect, he knew that whenever he looked the list over, he would think: Yes, I *want* to do all these things, they serve my interests.

And that's just what's lacking in no-e-mail Fridays. The name alone is negative—a prohibition of the very thing workers are addicted to—and it gives the whole concept a negative spin. It assumes that people are simply weary of their inboxes and would be thrilled to have a day off. In fact, the conundrum is more complicated than that. We love e-mail *and* we hate it. It lifts us up *and* knocks us down. By stressing only the negative, no-e-mail Friday gives short shrift to the very desire it's trying to combat and offers no positive goal to replace it. It's like naming a diet "The No Ice Cream or Any Other Goodies Diet." Who wants to go on that?

The other problem with today's workplace rituals is they tend to overlook the importance of inner conviction. It's not enough to tell employees that their screen habits are bad for them and hurting the company, and henceforth they shall adhere to the following rules. That's a recipe for failure. You have to give them a new way of looking at the problem that they can believe in.

One largely unrecognized downside of computers and other digital devices in the workplace is that they keep everyone relentlessly focused outward, beyond themselves. For people stuck in cubicles all day hooked up to screens, this sends an unfortunate message. The implication is that they're just conduits for data, with nothing valuable to offer in themselves. Obviously, in order to do their jobs effectively and contribute to their organizations, workers need information from the world at large. But to turn that information into ideas and initiatives of real value, they must bring their own unique talents and insights to bear. Rather than just aiming to modify employee *behavior* through e-mail prohibitions and the like, if companies focused on the *thinking* that drives the behavior, that alone would send a powerful message: what really matters is the untapped potential inside the employee, and the object of the ritual is to make the most of it. *Spending some time away from screens will bring out the best in you.*

Positive rituals based on inner belief—could they help workers with digital-dependency issues? There's evidence they could. One of the first large companies to recognize the threat to productivity posed by overload was the Intel Corporation, the world's largest manufacturer of the semiconductor chips that drive modern technologies. Intel has devoted unusual attention to this problem, experimenting for years with various strategies and techniques, including rituals. The company's experience was the subject of a recent study by the technology research firm Basex, which looked at several specific Intel programs aimed at getting workers to put some distance between themselves and their inboxes. It focused on three seven-month-long pilot programs whose subjects were Intel managers and engineers:

1. **Quiet Time** A weekly four-hour period in which the workers' incoming e-mail was shut down (they could compose and read e-mail, but not receive it), their instant-message status was set on "do not disturb," incoming phone calls were forwarded to voice mail, meetings were not scheduled, and signs were placed on office doors requesting privacy.

2. **No-E-Mail Day** On Fridays, whenever possible, the employees agreed to use verbal communication rather than e-mail. This was not a strict prohibition, but an effort to encourage person-to-person interaction within the group.

Outside e-mails were allowed, but members of the work group were discouraged from e-mailing each other unless necessary.

3. **E-Mail Service Level Agreement** The goal of this unfortunately named initiative was to lengthen the acceptable time period for replying to e-mail. Rather than feeling they must respond immediately to internal e-mails, workers could take as long as twenty-four hours to reply. It was hoped that, as a result, they would stop monitoring their inboxes constantly, and instead check just two or three times a day.

Quiet Time was the most successful of the pilots, winning the most positive reviews from participants and delivering stronger results (better concentration, more tasks finished on time, and so forth). When it was over, a plurality said they would like to continue the program. Though the study doesn't mention it, anyone familiar with the Franklin approach can't help but notice that Quiet Time was the only one of the initiatives with a positive name communicating an attractive goal. (An earlier Intel program that was considered a success had a similarly enticing name: YourTime.) While the other two of the more recent programs were considered failures and were discontinued, Quiet Time was extended beyond the pilot period, and at the time of the Basex study it was being evaluated for wider use in the company.

This is not to say that everyone liked Quiet Time, or that it's the answer. It was more popular with managers than with engineers, and not all subjects used the time to engage in the deep thinking for which it was intended. Some used it to organize and catch up on the contents of their (disconnected) e-mail inboxes. There were also problems with participants who not only didn't follow the rules themselves but also imposed on others who were trying to comply. As the study noted, for such programs to succeed, workers "must clearly communicate to others their availability and respect that of others." This is a reminder that, in the office and beyond, there are two distinct sides to this question: (1) My behavior affects *my* quality of life, and (2) My behavior can affect *your* quality of life.

Many Quiet Time participants complained that the underlying premise of the pilot, a *mandatory* period for reflection, was unrealistic given that people's jobs and needs vary greatly. Indeed, one of the frequent complaints about all company-wide limits on digital habits is that for some kinds of work, such as sales and customer service, e-mail and other screen applications really help get the job done, while for others, such as design and strategy, they often get in the way. Instead, individuals should be allowed to design their own rituals tailored to their specific situations. As one Quiet Time participant put it: "We should have at least four hours per day of uninterrupted time to work, and it shouldn't have to be a mandated program. People need to be more disciplined." In other words, the impetus should come from within.

Even if positive, inner-directed rituals like this took off, they wouldn't solve the root cause of office overload, which is the sheer ongoing growth of information and not enough time to handle it all. E-mail is only a piece of the puzzle. As Jonathan B. Spira, the CEO of Basex, has written, "Information overload is far more complex than too much e-mail." Nevertheless, if workers were reining in their own screen habits because they were convinced it was a good idea, it's hard to imagine that the situation wouldn't begin to improve. Instead of no-e-mail Fridays, a few hours of reading Ben Franklin might be more useful.

"All new tools require some practice before we can become expert in the use of them," Franklin once wrote. His discovery of how electricity works helped lay the groundwork for this electronic age. Given his strong crowd leanings, it's easy to imagine that, were he alive today, he'd be a massive screen addict. And in all likelihood he'd create a ritual for himself based on some notion of less screen time yielding more time for other highly desirable pursuits. Working on new inventions, perhaps, or his old favorite, venery.

Human nature hasn't changed much since the eighteenth century. Look inward first, and accentuate the positive. The ritual will write itself.

Chapter Ten

THE WALDEN ZONE

Thoreau on Making the Home a Refuge

"I had three chairs in my house; one for solitude, two for friendship, three for society."

I once read that someday the walls of the typical American kitchen will be constructed of enormous digital screens. The report had a sanguine tone, a perky world-of-tomorrow certitude that this will be a brilliant addition to any modern home.

Futurists have a remarkable knack for being dead wrong, but there were reasons to take this forecast seriously. There's no question that it's technologically feasible. Wall screens are increasingly common in public places, and some serious tech enthusiasts have had them at home for years. Less certain was the assumption that the broader public will welcome the chance to be bathed in floor-to-ceiling digital data as they chew their cornflakes. But based on consumer patterns of the last decade, including my own, it's not all that far-fetched.

I remember my excitement when I first heard years ago that a device was on its way that would distribute a broadband Internet connection wirelessly *all through the house*. This struck me as excellent news. We were still living in the city then and had two Internet-connected desktop computers, one in my home office and one in Martha's. So if for some reason we wanted to work on a laptop in, say, the kitchen, we had to tap in through a telephone jack, a cumbersome process. Rooms without jacks were "dead" zones unless you had a really long cord.

How nice it would be to have an effortless digital connection anywhere I wanted it in the house. Wherever the urge hit me, I could just "surf the Web," as we used to say, a phrase nicely evoking the adventure and personal freedom the burgeoning medium offered. Catch a digital wave, and you were sitting on top of the world. I saw myself happily surfing Amazon.com from a chaise in the backyard. When Wi-Fi routers duly arrived I paddled right out and bought one, and in no time we were a thoroughly connected household.

The surround-screen kitchen would simply take the same principle to a new level. Sure, in our wireless broadband home we can now go online from any room. But most laptop and smart phone screens are so small, the connected experience is inherently limited in scope. And to the digital maximalist, limits are the enemy. For instance, when I'm having a wireless connected experience at the kitchen table, as I sometimes do with my laptop, the material world often intrudes. If one of our cats happens by in my peripheral field of vision, I'm liable to pick it up and stroke it while burbling meaningless baby talk, losing my train of digital thought. Screen walls would diminish the wandering-cat effect. The digital sphere would more fully command the room and my attention. I don't relish this prospect, but some people apparently do.

Besides, the futurist thinking goes, there's something wonderfully elegant and Jetsons-esque about connected kitchen walls. Rather than looking into the electronic realm through a tiny window, we'd be living in and moving *through* it all the time. E-mail fonts could be a foot tall, while the life-sized people in videos would feel as though they were right there with us in the room. And imagine the convenience. If you suddenly needed a recipe or were curious about the overnight stock market numbers from Asia or wondering if the HR folks had replied to your last message or just wanted to wave to Grampa, you could reach out as easily as you do for the butter dish, touch a spot on the wall, or say a few words (these walls will have intelligent ears) and make it happen. Why stop at the kitchen? In some scenarios, the whole house will someday be a full-blown screen environment, every surface seamlessly digitized and world-fastened. And when that day finally arrives, we'll all be very . . .

Very what? What would it be like to live in such a thoroughly digital domicile? We don't know. Do we care that we don't know? Do we give it any thought amid the daily chatter about technology, the blithe upgrading to whatever is new and more connected? We think incessantly about the technology itself but not about how it's shaping everyday experience. And so, with our tacit permission, everything is becoming a digital "platform," even the home and, by extension, the people in it.

"Home" means so many things. On the most basic level it's simply a location, the place where one lives. It's also the physical structure, the house or apartment that is home. Last, home refers to the environment that's created inside that structure, a world-away-from-the-world offering refuge, safety, and happiness.

It's this last idea of the home as sanctuary that's absent from most thinking and decision making about technology. A kitchen with giant digital screens for walls would certainly offer convenience, but a household isn't just another utilitarian gadget. Like all connective devices through history, wherever screens go, they bring the crowd and the busyness that comes with it. This, in turn, has a powerful effect on how we think and feel. The home has traditionally been a shelter *from* the crowd, within which human beings experienced life in a different way from how it was experienced on the outside. For the individual, home has always offered privacy, quiet, solitude. For those living in couples, families, and other cohabiting groups, it also afforded an intimate sort of togetherness that's possible only in shared isolation.

The crowd drives us away from the reflective, the particular, and the truly personal. At home we could be more human.

High-speed, around-the-clock digital connectedness has already diluted these vital aspects of home life. The more connected our house became in the last decade, the less it provided the sense of peace and soul nourishment I associate with "home." What was once a happy refuge from the crowd is becoming a channel for crowd delivery. The walls are membranes through which a tide of people and information flows in and out around the clock. It's not just online friends, interests, and work duties but news, popular culture, and the never-ending bustle of the marketplace. We're swimming in massacres and tragedies, drowning in celebrities, trends, fads, sensations, crazes. It sucks you in, and as it does, the here-and-now experiences and interactions that should be the core of home life are reduced to fading background music.

I didn't see this coming. Radio and television have been delivering their own crowds into private homes for generations, and the telephone has long been a link to the world at large. I must have been assuming in some unconscious way that always-on digital connectedness would be more of the same. In the long run, it may be. Perhaps it's just the newness of it that makes the interactive screen experience seem so much more intense than what older technologies offer. Just a few generations ago, television was viewed as an invasion of the sacred space that is home and a particular menace to children. Those dangers are still real today, but over the years it's also become clear that, if used properly, television can be a useful tool as well as a gathering place, an alternative hearth. Television is just that in our house, where we carefully regulate its use and enjoy it immensely. So we may just be at the start of an adjustment period, and someday it will seem silly that anyone ever questioned the wisdom of that universally beloved instrument of happiness, the digitally walled kitchen.

But you can't live in the future. In the reality that is the present, these devices have a mesmerizing hold on us, and it's altering the nature and meaning of our domestic lives. One of the most reliable routes to inwardness and depth has become an increasingly outward experience. How can you relax and recharge when *the whole world* is living with you?

We're already pretty far down this road, and the question is whether it's still possible to do anything. Can this drastic repurposing of the home be amended or modified so it remains a home in every sense of the word?

I think so, and the best way to see how is to go back to the origins of today's wired world a century and a half ago and the unlikeliest of all digital philosophers, Henry David Thoreau. In the familiar telling of his story, Thoreau would seem to be the last person with anything useful to say about managing home life in a digital world. He's best known for abandoning civilization for the one-room house he built in the woods outside Concord, Massachusetts, where he lived a simple life close to nature. *Walden*, his account of that experience, is ostensibly a rejection of society and the insidious ways it warps us and robs life of its richness. In making his case, he often mentions technology, in particular two new inventions that were transforming the world, the railroad and the telegraph.

At a time of rapidly growing connectedness, Thoreau disconnected. He was the great escape artist, and escape would seem to be his message. If you want to take back your life, Get out! Or, as he puts it in *Walden*:

> I went to the woods because I wished to live deliberately, to front only the essential facts of life, and see if I could not learn what it had to teach, and not, when I came to die, discover that I had not lived. . . . I wanted to live deep and suck out all the marrow of life.

The essential problem hasn't changed, nor has the goal. Who doesn't want to live the fullest, deepest life they possibly can? For the overconnected soul wishing to apply Thoreau's message, however, the sticking point is his method. As a practical matter, not many people have the freedom to escape society—jobs, family, and other obligations—and hole up in the woods. In any case, very few of us want the pure solitude that Thoreau seems to be advocating when he writes, "I love to be alone. I never found the companion that was so companionable as solitude."

It's the rare person whose ideal of home is a cabin for one in a neighborhood without neighbors. Part of what's always been special and invigorating about the typical home is that it makes solitude available *within* the context of the larger social environment. It's an intermittent respite, a space into which one retreats briefly at regular intervals, to emerge later refreshed.

Today there's another factor that makes Thoreau's approach seem not just unappealing but downright pointless. Even if we wanted to run away physically from society, in a digital world there's no place to go. With ubiquitous mobile connectivity, you can't use geography to escape what he called society, because it's everywhere. If you have a screen of any kind with you—and who doesn't these days?—you haven't left society at all.

But to dismiss Thoreau for these reasons is to miss the whole point of *Walden* and its relevance to our time. In fact, he wasn't trying to escape civilization, and what he created at Walden Pond was not even close to pure solitude. As for ubiquitous technology, it's true that the world was a lot less connected in the middle of the nineteenth century than it is today. However, Thoreau lived through a major technological shift, the arrival of instant communication, that foreshadowed the current one. The woods weren't wireless in his era, but for the first time in history they were getting *wired*, and the wires were carrying information around the world at unimaginable speeds. Thoreau saw the enormous human implications of this change, and he structured the Walden experiment so that it spoke not just to his own time but to the technological future he saw coming.

In a world where it's increasingly hard to escape the crowd, can you still build a refuge, a place to go inward and reclaim all the things that a too-busy life takes away? Thoreau says you can, and he offers a practical construct for making it happen. *Walden* can serve as a philosophical guidebook to the tricky challenges of twenty-first-century domestic life, including the matter of connected kitchen walls. The quickest way "home" in a digital world is to follow Thoreau.

FIRST, HOWEVER, IT'S necessary to correct a few misimpressions, beginning with the idea that Thoreau was trying to escape society. Walden Pond was not exactly Antarctica. It was just a mile and a quarter from the town of Concord, where Thoreau grew up and spent nearly all of his life after college. To him, the world of Concord *was* society in the most immediate sense, and when he speaks in *Walden* about the harried lives of his contemporaries—"The mass of men lead lives of quiet desperation"—he was thinking especially about his friends and neighbors. He refers to them often, collectively and individually; they are his chief source of real-life examples, the evidence for his diagnosis of society's ills. Despite the hell-is-other-people statements he sometimes tossed off, he saw those people frequently while at Walden. For a famous recluse, he had an unusually active social life, which he describes in a chapter called "Visitors." Though the cabin was only ten by fifteen feet, he entertained as many thirty people there at one time—hardly a hermit's life.

Moreover, when the twenty-seven-year-old Henry moved to Walden in the summer of 1845, the railroad, which was society in motion, came with him. A brand-new track had just been laid connecting Concord to Boston and the rest of the world, and it ran right beside the pond. He could see and hear the trains from his place. The railroad wasn't just a visual and aural symbol of civilization, it was a dynamic reminder of how technology was making the world much smaller in the middle of the nineteenth century. In today's terms, it would be like building your rustic retreat in the woods beside the runaway of an international airport. If he really wanted to escape society, Thoreau could have done much better. He liked to go on wilderness trips around New England and certainly knew more remote places.

He went to Walden because that's where the opportunity presented itself. The owner of the land was Ralph Waldo Emerson, the Concord philosopher who was his mentor and friend, and the location made practical sense in a number of ways. He was going to be very busy in this endeavor, writing, growing vegetables for both meals and income, working at other odd jobs to support himself, and keeping house. For logistical reasons it would be far easier to do all this close to town, where he knew everyone and where there were stores, a post office, and other conveniences.

Beyond these practical considerations, the fact that Walden was so close to Concord was a key element of the venture, crucial to its meaning and value. He'd recently spent the better part of a year living on Staten Island, where he had been unhappy. He'd "learned that his heart really was in Concord," writes his biographer Robert D. Richardson, Jr., and the rest of his life would be firmly grounded there. This was home, in other words, and in *Walden* and his other writing projects he was consciously exploring the meaning of that home, as well as "home" in a more general sense: What is a home, really? What kind of home makes us happy?

Walden isn't just a philosophical tract, it's a detailed account of one man's life at home, from the nitty-gritty economic details—he provides elaborate charts of household expenses and revenues—to the spiritual and emotional experiences that living there yielded. This wasn't merely a shelter, it was a place to "live deep," as all the best homes are. Thoreau had times of intense happiness, even ecstasy, in his home, and they're central to the book's message.

His nearness to society also made the project relevant to others. If he *had* fled to a truly remote place, his life there would have borne no resemblance to the lives of most other people, and they'd be unable to emulate it. "It would be some advantage to live a primitive and frontier life," he wrote, "though in the midst of an outward civilization." That is, he consciously didn't flee the busy world of society but instead set up camp just on its periphery. "Thus," Richardson points out, "it was clear to him at the very outset that what he was doing could be done anywhere, by anyone. It did not require a retreat from society. . . . He himself thought of it as a step forward, a liberation, a new beginning, or as he put it in the second chapter of *Walden*, an awakening to what is real and important in life." An awakening that others could have in their own homes, if they wanted it.

But can we apply *Walden* to our time? Thoreau may have been close to town, but he wasn't holed up with the rest of the planet, as we are with our screens. Given that digital technology has so altered the landscape of modern life, and particularly life at home, is it a stretch to think Thoreau could have anything useful to say to us?

Not at all. Though it's true that he lived in a very different information environment from today's, he and his friends and neighbors really *were* living close to the rest of the planet in a new way. Previously, information could travel only as quickly as the swiftest mode of physical transportation, which was trains. With the arrival of the telegraph in the 1840s, messages could suddenly dart from place to place instantaneously. Oceans, deserts, and mountain ranges were no longer barriers. All it took was a wire. The notion that one could now theoretically keep up with anything and everything happening on Earth, and around the clock, was both thrilling and unsettling. An East Coast American of Thoreau's generation wasn't just increasingly connected to the wide world, he was increasingly immersed in it, and he needed to manage that immersion. What to read? What to care about?

This was a subtle but significant shift in the nature of inward life, and everyone was grappling with it. "A slender wire has become the highway of thought," observed the *New York Times* in an editorial published on September 14, 1852.

Messages follow each other in quick succession. Joy spreads on the track of sorrow. The arrival of a ship, news of a revolution, or a battle, the price of pork, the state of foreign and domestic markets, missives of love, the progress of courts, the success or discomfiture of disease, the result of elections, and an innumerable host of social, political and commercial details, all chase each other over the slender and unconscious wires.

With a little updating of the language, this could be a description of the moment-by-moment randomness now offered by any digital screen. There was simply a great deal more information bearing down on everyone, and even the home was no safe haven. In *The Victorian Internet*, a history of the telegraph, Tom Standage quotes W. E. Dodge, a prominent telegraph-era businessman from New York, describing the plight of a family man battling information overload:

> The merchant goes home after a day of hard work and excitement to a late dinner, trying amid the family circle to forget business, when he is interrupted by a telegram from London, directing, perhaps, the purchase in San Francisco of 20,000 barrels of flour, and the poor man must dispatch his dinner as hurriedly as possible in order to send off his message to California. The businessman of the present day must be continually on the jump.

In other words, the telegraph was the latest agent of the "quiet desperation" that Thoreau saw all around him and felt in himself. Devices meant to relieve burdens were imposing new ones, pulling people away from life's most meaningful experiences, including the family dinner table. "But lo! men have become the tools of their tools," he wrote, and though he wasn't specifically referring to the telegraph, elsewhere in *Walden* he made it clear that the slender wire could make tools out of people. New technologies, he said, are often just "pretty toys, which distract our attention from serious things. . . . We are in great haste to construct a magnetic telegraph from Maine to Texas; but Maine and Texas, it may be, have nothing important to communicate." Yet at other times, he wrote about the telegraph in a hopeful, lyrical way, suggesting he saw the wonder of the technology and perhaps its potential to do good. "As I went under the new telegraph wire, I heard it vibrating like a harp high overhead," he noted in his journal. "It was as the sound of a far-off glorious life."

Naturalist that he was, it's often assumed that Thoreau loathed technology. In fact he was a sophisticated user, and occasionally a designer, of technologies. He never made much money from writing and supported himself by working in two different tool-intensive fields: as a surveyor and in the pencil-manufacturing business owned by his family. At one point he took on the ambitious project of reengineering the Thoreau pencil so it might fare better in a competitive marketplace. He worked hard on it, conducting extensive research into why certain European-made pencils were so superior to their American counterparts. Based on what he learned, he changed the materials, design, and manufacturing process of his company's pencils, essentially developing a brand-new product. His efforts were a great success, producing "the very best lead pencils manufactured in America" at the time, according to Henry Petroski's *The Pencil*, a history of the tool.

Thoughtful student of technology that he was, Thoreau saw that as the latest connective devices extended their reach into the lives of individuals, they were exacting huge costs. They're the same costs we're paying today—extreme busyness and a consequent loss of depth. The more wired people became, the more likely they were to fill up their minds with junk and trivia. What if we built this fabulous global telegraph network, he wondered, and then used it only to keep up on gossip about *celebrities*? "We

are eager to tunnel under the Atlantic and bring the Old World
some weeks nearer to the New; but perchance the first news that
will leak through into the broad, flapping American ear will be
that the Princess Adelaide has the whooping cough. After all,
the man whose horse trots a mile a minute does not carry the
most important messages."

That is, he saw that instant communication had the potential to exacerbate the very problem he had gone to Walden
to solve, the superficial, short-attention-span approach to life
that afflicted his friends and neighbors and often himself. They
were all living from one emergency to the next, he writes at
one point, consumed by their work, always checking the latest
news. "Why should we live with such hurry and waste of life?
. . . We have the Saint Vitus' dance, and cannot possibly keep
our heads still." Saint Vitus' dance is a nervous disorder whose
symptoms include sudden, jerky movements of the limbs and
face. The name comes from a mysterious social phenomenon
first observed in Aachen, Germany (the city of the little mirrors), in the fourteenth century, when large numbers of people
simultaneously broke out into wild fits of frenzied dancing,
foaming at the mouth in some cases. Now the weird dance was
in the mind.

Once the consciousness was hooked on busyness and external stimuli, Thoreau saw, it was hard to break the habit. Never
mind the telegraph, even the post office could become an addiction, as he observed in a speech:

Surface meets surface. When our life ceases to be
inward and private, conversation degenerates into mere
gossip. . . . In proportion as our inward life fails, we
go more constantly and desperately to the post-office.
You may depend on it, that the poor fellow who walks
away with the greatest number of letters proud of his

extensive correspondence has not heard from himself this long while.

This is the problem of *our* time, too, of course. And it's what he went to *Walden* to solve. The mission: to see if, by building a home at a slight distance from society—disconnected, yet still connected in many ways—and living there thoughtfully, he could go back inward, regaining the depth and joy that was being leached out of everyday life.

Among all those who were struggling with this challenge in the mid–nineteenth century, Thoreau was unusually well situated to find an answer. Concord was the center of American Transcendentalism, a philosophical movement that provided a rich vein of pertinent ideas. Transcendentalists believed that true enlightenment does not come from other people or outward sources such as organized religion, scientific observation, and books; rather, it comes from within. The profoundest truths about existence are available to each of us through intuition and reflection.

It was a philosophy that spoke directly to a time when trains and telegraph lines, as well as industrialization and other forces of modernity, were pulling people in exactly the opposite direction—outward. The crowd seemed terribly important and powerful in those days, just as it does now, and it was hard to resist its influence. It was as if you had no choice but to submit, fall in line. The Transcendentalists believed that resistance was crucial. Emerson, the movement's leading figure, wrote in his great essay "Self-Reliance" that to be truly happy and productive, you have to tune out the crowd and listen to "the voices which we hear in solitude." In another piece, Emerson described a Transcendentalist as a person who essentially wakes up one day and realizes, "My life is superficial, takes no root in the deep world." And then does something about it.

Guided by this philosophy, the Walden project was really an exercise in practical reengineering. In this case, the device that needed redesigning wasn't a pencil but life itself. Thoreau's method was to strip away the layers of complexity that outer life imposes, to "Simplify, simplify," as he wrote, and, in so doing, recover that lost depth. As Thoreau scholar Bradley P. Dean puts it, "By simplifying our outward lives, we are freer and better able to expand and enrich our inward lives."

The heart of the effort, serving as both headquarters and object lesson, was Thoreau's tiny house and the life he constructed there. It was seriously spartan, reflecting the simplicity creed. But there was another kind of simplicity that mattered even more than the material kind: simplicity of the mind. Though the house was right in the midst of civilization, close to town, in sight of the railroad, and within easy reach of visitors, he defined it as a *zone* of inwardness, and that's what it became.

In effect he put up invisible philosophical walls that said: No news, busyness or stimulation, including the human kind, enters here without my permission. There were visitors, absolutely, and he welcomed them. "I love society as much as most, and am ready enough to fasten myself like a bloodsucker . . . to any full-blooded man that comes in my way." But they came intermittently, and generally for good reason. In town, people would drop by on any excuse, but here "fewer came to see me on trivial business. In this respect, my company was winnowed by my mere distance from town." However, distance wasn't the only factor. This space had been zoned for a purpose, and people knew it, or they found out. When they overstayed their welcome, he let them know: "I went about my business again, answering them from greater and greater remoteness." Thus, the crowd was never overwhelming. There was space and time to be alone, and with others—a healthy human mix. "I had three chairs in my house; one for solitude, two for friendship, three for society."

Walden recounts an experiment in building a good home by adopting a new idea of what home is all about and living by it. "So easy is it," he writes, "though many housekeepers doubt it, to establish new and better customs in the place of the old." It was a successful experiment: Thoreau had the spiritual awakening he'd hoped for, and it's reflected on every page.

> It is something to be able to paint a particular picture, or to carve a statue, and so to make a few objects beautiful; but it is far more glorious to carve and paint the very atmosphere and medium through which we look, which morally we can do. To affect the quality of the day, that is the highest of arts.

Reading him has produced a similar effect on generations of people around the world, and occasionally on history. Among the countless people influenced by Thoreau was Gandhi, who cited him as a major inspiration of his own philosophy and the Indian independence movement.

And because it was an experiment conducted in easy reach of society, in what Robert Richardson calls "a backyard laboratory," it can be replicated in any home. *Walden* shows that, even in the midst of a frenetic world, one can create a zone where simplicity and inwardness reign—a sanctuary from the crowd. The need is far more pressing now. Thoreau reports that many of his visitors were mystified by his project, didn't see the point. Today, a zoned-off sanctuary for the heart and mind couldn't make more sense. It's why we go to spas and yoga classes, *leaving* home to obtain what used to be home's special gift.

There's a long-standing awareness in architecture and design of the value of domestic zoning, to ensure that a home serves all the needs of those who live there. In the late 1930s, an influential book called *The Human House* by Dorothy J. Field made the case that every house should have zones designated for various degrees of solitude and togetherness, privacy and activity. In other words, a house should offer its occupants the opportunity to move back and forth along the continuum of connectedness. Focusing on the family dwelling, Field wrote, "All thoroughly satisfactory family houses *are* zoned. In any such house you can find a room which is always a quiet room, a room where you can always enjoy a romp, noise, or any activity without shushing or nagging, and a private cubbyhole for yourself to retire to." Her ideas influenced Frank Lloyd Wright and other thinkers.

Zoning is way overdue for a comeback, a digital revival, and it's surprising it hasn't happened yet. Thoreau could be the model. Our situation is different from his, in that the crowd is no longer just nearby—it's right in the home, wherever there's a screen. So our zoning has to be interior. Every home could have at least one Walden Zone, a room where no screens of any kind are allowed. Households that take their tranquillity seriously, and have sufficient room, might designate such a space for each person. There could be a shelf or cabinet outside the doorway where, upon entering, all smart phones and laptops are turned off and put away.

The wireless signals in those rooms won't go away, of course, and that's a problem. But as with Thoreau, the point of the zone is to use an idea as a constraint on behavior. For a Walden Zone to work, you first have to *believe* it's a good idea; once you do, it's a lot easier to resist temptation. The mind puts up an invisible wall, which blocks the invisible signal. Technology could help, too. Perhaps a canny entrepreneur with an eye to the Thoreauvian future will come up with a device that scrambles wireless signals in any designated space.

The opposite of a Walden Zone would be a Crowd Zone, any room specifically designated for screen life. Home offices would be automatic Crowd Zones for most people. Since the kitchen is a natural gathering place in many homes, it's a good Crowd Zone candidate. In a thoughtfully zoned house, a kitchen with floor-to-ceiling wall screens begins to make sense. Connectedness is much more appealing and rewarding when you know there's a place nearby to get away to.

Another option is whole-house zoning, in which the entire dwelling becomes a Walden Zone during certain times of the day or certain days of the week. This requires more commitment, as it means truly swearing off screens during designated times. The advantage of this approach is that it creates a genuine refuge, as Thoreau's house must have been on quiet winter nights when the town seemed a thousand miles away. My family has had great success with a regimen of this kind, which I describe in detail in part III.

The point is not to withdraw *from* the world but *within* the world. It's funny that Thoreau, of all people, should be the source of this wisdom. But remember, Walden was just a two-year experiment. When it was done, he returned to society and lived the rest of his life there. But he took a valuable piece of knowledge with him: you *can* go home again, whenever you need sanctuary, so long as you have a home that serves this purpose. It doesn't have to be far off in the woods or up in the mountains or anywhere special. It's not the place that matters, it's the philosophy. To be happy in the crowd, everyone needs a little Walden.

"You think that I am impoverishing myself by withdrawing from men," Thoreau once wrote in his journal, "but in my solitude I have woven for myself a silken web or *chrysalis*, and, nymph-like, shall ere long burst forth a more perfect creature, fitted for a higher society."

Chapter Eleven

A COOLER SELF

McLuhan and the Thermostat of Happiness

"How are we to get out of the maelstrom created by our own ingenuity?"

At the end of an e-mail, a friend mentions how crazy her life has become, particularly at the office. She has a good job at a prestigious university, the kind of place I like to imagine as being somewhat insulated from the chaos. I ask what she means by crazy.

"The i.m.'ing knows no bounds," she replies. "It feels like my central nervous system is interlinked with all my colleagues'."

A brief description, less than twenty words. But I know just what she's talking about, and she knows that I know. We're both interlinked to more people than it's possible to hold in the mind at one time. Everyone is. And the panicky feeling behind her words, the sense of being plugged into an infinite crowd from which there's no unplugging, is the characteristic sensation of this era.

So far, the ideas explored in this part of the book have come from the distant past. There have been all kinds of striking parallels between past and present, ways in which people of previous eras felt much as we feel now. But the fact is, none of them experienced exactly what we're experiencing.

Thoreau walked under telegraph wires and heard them sing, but he never watched an event unfold on the other side of the world in real time. He never typed in a search term and instantly got back 25 million results. He never woke up in the morning to find that 150 new messages had arrived overnight, silently, inside a waferlike object glowing on the nightstand—an object that really *does* seem to have a direct line into the nervous system. Yet there is a way to pull the good ideas from the past into the reality that surrounds us. Marshall McLuhan, the only philosopher in this survey who lived in the age of screens, provides the missing piece.

Today McLuhan is known primarily for two catchphrases that he coined, "the global village" and "The medium is the message." They weren't just slogans, they were prophecies, and astonishingly good ones. He saw this digital world of ours coming, and he wrote a great deal more about it than two phrases. He left behind a sprawling, penetrating, idiosyncratic body of work, a whole philosophy aimed at making sense of life in a world made much smaller and busier by electronic technology. His overriding theme was that, even in a hyperconnected world, everyone has the ability to regulate his or her own experience.

At the time it was widely feared that mass media were turning people into helpless automatons. The crowd was on the rise again, and McLuhan wanted people to know that if they felt overwhelmed by technology—"involuntarily altered in their inmost lives," as he put it—it didn't have to be that way. They could take control of the situation, just by living more consciously.

It's the same theme that great thinkers have struck time after time over the last two thousand years, but it keeps getting forgotten. The answer to our dilemma is hiding in the last place we tend to look: our own minds. McLuhan believed that even at a time when technology, and the crowd it delivers, has direct access to the mind, the best tool for fighting back is *still* the mind itself. His mission was to update the mind's arsenal for the new challenges of the future. That future has come, and though McLuhan died thirty years ago, his message couldn't be more timely.

McLuhan was a Canadian academic, a scholar of English literature with a passionate interest in mass media and popular culture. In his early writings, he examined the content of media, particularly advertising. At the time, this was the standard way of thinking about technology: it was ideas and messages that mattered, not the devices that delivered them.

This is not to say the technology was ignored. Radio and television pulled huge new crowds together in the 1950s and early '60s, giving birth to mass society, and there was enormous concern that individuals were losing the ability to think for themselves. It was in 1950 that sociologist David Riesman's book *The Lonely Crowd* came out, garnering wide attention with its argument that human beings were becoming less "inner-directed," or guided by their own values and beliefs, and more "other-directed," or shaped by those of society. Outwardness was replacing inwardness.

Numerous other books and movies of this period grappled with the meaning of the crowd and its impact on people's minds and behavior. Some, such as *The Organization Man* and *The Man in the Gray Flannel Suit*, focused on the soul-killing conformity of life in the corporate world. Others saw demagoguery in politics as a growing threat. World War II was a fresh memory, and Hitler and other fascist leaders had been skillful manipulators of mass opinion. The fear was that new rabble-rousers could use the electronic media to spread poisonous messages. In his influential book *The True Believer*, a San Francisco longshoreman turned philosopher named Eric Hoffer examined why individuals willingly surrender their freedom and individuality to mass movements. In the 1957 movie *A Face in the Crowd*, Andy Griffith played a simpleminded country singer who becomes a media celebrity and political demagogue. But it was the messages themselves, and the charismatic personalities of the people sending them, that were thought to be the real source of power. Technology was basically a conduit.

Meanwhile, the burdens of this new life in the crowd were being felt in the ordinary comings and goings of daily existence. In *Gift from the Sea*, Anne Morrow Lindbergh wrote in 1955 about the crushing load of obligations bearing down on the modern self:

> For life today in America is based on the premise of ever-widening circles of contact and communication. It involves not only family demands, but community demands, national demands, international demands on the good citizen, through social and cultural pressures, through newspapers, magazines, radio programs, political drives, charitable appeals and so on. My mind reels with it. . . . It does not bring grace; it destroys the soul.

Lindbergh's book reads like a prequel to the digital age. Connectedness had ramped up dramatically, and, as today, it was turning life into a slog. But notice that she, too, focused on the content of her busyness—the various "demands" that arrive "through" media and other sources—rather than the technologies themselves.

Thus, on two different planes—the macro (sociopolitical life) and the micro (private life)—there was a broad sense that, in an increasingly crowded world, people were less free to be themselves. Whether they were surrendering their minds to a charismatic ideologue on the radio or simply unable to keep up with everyday demands and distractions, the effect was the same: they were losing their autonomy, becoming creatures of the outward world. And, the thinking went, this was all the result of incoming messages and ideas, of content.

Few stopped to consider the gadgets that connected everyone in the first place—radio, television, and so on—and what role those might be playing, quite apart from the content they conveyed. That's where McLuhan came in. In 1962, with his groundbreaking book *The Gutenberg Galaxy: The Making of Typographic Man*, he proposed a completely new way of thinking about this question. He argued that the technologies themselves have a more powerful impact on human beings than the content they carry. This, he explained, is because our tools are really extensions of our bodies.

Written language, for example, is an extension of our sense of sight: it extends our vision out into the world, allowing us to pull back information in the form of letters and words. Whenever a new connective device is added to the toolbox, it extends another part of us outward. The telephone gave our ears global reach, while television extended both eyes and ears in a new way. According to McLuhan, every time this happens it alters how we perceive and process reality, in effect creating a new environment for the mind and for our lives. We inhabit a reality shaped fundamentally by our tools. Thus the medium is the message, far more than the content it carries.

This outward extension of the self through tools had been going on since the dawn of human history, and because it involved a fundamental rearrangement of one's mental life—the old expression "rock my world" neatly sums up the effect—it was always stressful. "Man the tool-making animal, whether in speech or in writing or in radio, has long been engaged in extending one or another of his sense organs in such a manner as to disturb all of his other senses and faculties," McLuhan wrote.

He took it a step further, contending that when a truly momentous new device appears, such as the printing press, the inner environmental change is so dramatic that it produces a new kind of human being. So, in addition to the medium being the message, *the user is the content.* We ourselves are changed by our devices, and because we're changed, society changes, too. Gutenberg's invention had created what McLuhan called Typographic Man, whose mind operated in a linear, objective fashion that fostered individualism. Equipped with this left-brain way of thinking, this being had thrived for centuries and built up Western civilization.

But McLuhan said he was about to be replaced. Because mass electronic media work on us in a different way from print, those technologies were creating a new person whose mind was less linear and individualistic, more group-oriented. In the future, he predicted, our minds would operate more like the oral mind of Socrates' era. In fact, he said, this new age had already arrived, which was why everyone was feeling so anxious and full of doubt. Print had given human beings the "inner direction" that Riesman had talked about in *The Lonely Crowd*, and now they felt it slipping away. The old boundary between the inward self and the outward world had been permanently breached by electronic technology. Inner direction would now be much harder to come by.

He traced this shift back to the nineteenth century, when, he said, the telegraph had, in effect, extended the entire central nervous system, including the brain, out into the world.

Suddenly, human beings were immersed in what he called "a total field of interacting events in which all men participate," i.e., everything happening at any given moment on the planet. By the mid–twentieth century, telephones, radio, and television had made this brain-taxing environment all the more intense. According to McLuhan, this was the true source of the stress and unhappiness people were feeling, the sense of the mind being under siege and paralyzed. His biographer W. Terrence Gordon summarized McLuhan's view: "Technologies create new environments, the new environments create pain, and the body's nervous system shuts down to block the pain."

However, there was a way to avoid the pain and thrive in the global village. McLuhan said it was a matter of understanding that you were living in this new world and then adjusting to it. Though he believed that the new gadgets were the source of our trouble, he didn't *blame* them. He placed the ultimate responsibility with human beings. If our technologies are driving us nuts, it's our fault for not paying attention to what they're doing to us. Why should we allow tools that are supposed to be making us happy to make us miserable? We should take control of the new technologies "instead of being pushed around by them."

His next book opened with the motto "The medium is the message," and it made him an authentic pop-culture icon. It was an unlikely fate for a fifty-two-year-old brainiac given to quoting James Joyce and Charles Baudelaire. But it was a moment when people were desperate to make sense of a crowded world, and he offered a fresh approach. To promote it, he deftly used the technologies he wrote about, appearing widely in the media, including on TV talk shows. Sometimes he would discuss his theories, but often he was just another famous-for-being-famous celebrity. On the kooky comedy show *Rowan & Martin's Laugh-In*, the question "Marshall McLuhan, what are ya' doin'?" became a running gag.

Unfortunately, though his catchphrases caught on, most people never fully grasped the concepts behind them. And that was really McLuhan's fault. His writing was too theoretical and maddeningly circular. His books were structured as collections of short stand-alone essays presented in what he called "mosaic" fashion, meaning they could be read in any order. It was his effort to break out of the linear thinking he believed was a thing of the past. For readers raised in a left-brain culture, however, it wasn't a helpful approach, particularly since it was delivered in a medium designed to be read from start to finish, the book. The abstruseness of his work eventually became part of his shtick and the theme of a funny moment in the Woody Allen movie *Annie Hall*, in which McLuhan plays himself. Even today, with the global village in full swing, reading him, one often feels like Alice in Wonderland trying to decode a barrage of seemingly random statements.

He was not a neuroscientist, and when he tries to describe the workings of the central nervous system, his language can be particularly inscrutable: "My suggestion is that cultural ecology has a reasonably stable base in the human sensorium, and that any extension of the sensorium by technological dilation has a quite appreciable effect in setting up new ratios or proportions among all the senses." If he'd written more plainly, his theories might be as well known today as his maxims.

Despite these obstacles, McLuhan has endured into this century, for a couple of reasons. First, in the early digital years his work was rediscovered and embraced by fervent fans of the new gadgets, who translated "The medium is the message" to mean "Technology rules!"—the exact opposite of how McLuhan believed the world should work. But this leads to the second reason he's endured and why he's so relevant today:

he placed human freedom and happiness before technology. Though our devices do have a tremendous influence on us, *we* should rule.

In his efforts to encourage this, he hit on some important truths about life in an electronic society. Sometimes it really does feel as though your brain is extended so far into the outward world, it's left your body. When this happens, it's very difficult to go back inward and be alone with your thoughts. That's what depth comes down to, really, taking all the stuff your mind has gathered in its travels back inside, to sort through it and see what it all means. To make it your own. The only way to cultivate a happy inner life is to spend time there, and that's impossible when you're constantly attending to the latest distraction. Attention deficit issues, Internet addiction, and other tech-related maladies are all about being stuck in outward gear.

McLuhan's prescription? He insisted he wasn't advocating any particular approach and didn't provide specific instructions about how to apply his work. His ideas are best if used selectively; his most valuable insight is that even though technology is impinging more than ever on our minds, they're still *our* minds. You can allow yourself to be led around by technology, or you can take control of your consciousness and thereby your life. He had a way with metaphors, and he had one for each of these options.

He used the Greek myth of Narcissus to explain why people become entranced by tech gadgets. Narcissus is the youth who sees his own reflection in the water and mistakes it for somebody else. "Now the point of this myth," McLuhan wrote, "is the fact that men at once become fascinated by any extension of themselves in any material other than themselves." Similarly, he said, we're fascinated by new technologies because they project us beyond ourselves. But just like Narcissus, we don't recognize that that's what the gadget is doing, project-ing *us*, by extending our bodies into the world. The confusion induces a kind of trance. We can't take our eyes off it, but we don't understand why.

His nickname for the Narcissus type applies to anyone who has ever been mysteriously spellbound by a screen (i.e., just about everyone): the Gadget Lover. But some people have it really bad. The cure, McLuhan said, "is simply in knowing that the spell can occur immediately upon contact, as in the first bars of a melody." Feeling the need yet again to stare long-ingly into the screen? Think of Narcissus and resist.

The second metaphor is about the active, take-charge approach he favored. Being hooked up to the crowd all the time—our central nervous systems "interlinked," as my friend put it—doesn't mean we have to surrender our fate to it. To make this point, McLuhan used an Edgar Allan Poe story called "A Descent into the Maelstrom," about a fisherman whose boat is sucked into an enormous whirlpool. Down he spins into the roaring vortex, certain he's going to die. Then something strange happens. In his delirium, he relaxes and, to amuse himself, makes a game out of studying how the whirl-pool works.

Other boats have been drawn in and demolished, and he notices that the pieces of flotsam flying past behave in differ-ent ways depending on their shapes. While most hurtle rapidly downward, cylindrical objects such as barrels aren't swallowed up as easily. Those linger up near the top of the vortex, closer to the surface. Based on this observation, he decides to lash himself to his own water cask and leap overboard. It works. The boat continues whirling down to its doom, but the clever fisherman doesn't. "The cask to which I was attached sank very little farther," he says. Eventually the whirlpool stops whirling and he winds up back on the surface. "The sky was clear, the winds had gone down, and the full moon was setting radiantly in the west." He's saved himself.

To McLuhan, the whirlpool stood for life in an electronic world. Here we are, surrounded by a ferocious, disorienting barrage of information and stimuli, seemingly spinning out of control. "How are we to get out of the maelstrom created by our own ingenuity?" he asked. His answer was to do what the fisherman did. Instead of panicking, take a deep breath and be resourceful. Study the flotsam of the moment, and grab onto something solid.

Poe's story was a favorite of McLuhan because, like his philosophy, it came down to the individual. Human ingenuity may have created our whirlpool, but it can also save us, one person at a time. We shouldn't be paralyzed by the new environment in which we find ourselves, but engaged and creative. "People are cowed by technology," Kevin McMahon, the director of a documentary film about the philosopher called *McLuhan's Wake*, once observed. "The optimistic side of McLuhan's message is: You've built these things, and you can control them if you understand how they affect you. To me, his message is still really important."

The logical question, then, is: what's *our* water cask? Like the fisherman, everyone has to work that out for himself. We're all different, and there's no one-size-fits-all way to balance the outward life and the inward one. That has always been true. What matters most is engagement, being conscious that you're shaping your own experience every moment. If you spend most of your time pressing keys and managing electronic traffic, that's what your life will be about. Maybe that makes you happy. If not, you have other options.

One helpful McLuhan technique, a refined version of his cure for the Narcissus trance, is to bear in mind that different devices affect us in *different ways*. To illustrate how this works, he used temperature as metaphor, distinguishing between "hot" and "cool" technologies. A hot technology is intense, overwhelming us with information and stimuli. A cool one is less intense, inviting the user to participate more in the experience, fill in the blanks. "The hot form excludes, and the cool one includes," he wrote.

He defined radio as a hot medium, because it intensely floods one sense with information, leaving little for the listener to fill in. But he said that television is cool, because it seeks more viewer involvement. These definitions are flexible and can change over time, since new technologies alter how older ones affect us. Today, though digital screens are highly participatory, they're also overwhelming and, arguably, hot. And radio now seems relatively cool.

The point is that by bearing in mind that gadgets have different effects, you can regulate the climate in your mind. It's another way of thinking about the continuum of connectedness that we're always moving along. If six straight hours of screen time has your mind overheated, what will cool it down? Staring at your handheld the entire subway ride home might not do it. Maybe it's better to just sit quietly and enjoy the ride. Sometimes the coolest device of all is no device. Rather than allowing external forces to define how we feel inside, each of us can be our own thermostat.

As instructive as McLuhan's ideas remain, even more remarkable in retrospect is how eager the world was to hear them. Half a century ago, there was enough interest in the human dilemmas posed by technology to turn an obscure literature professor into an international celebrity. And by raising consciousness about these questions, McLuhan further broadened that interest. For a time, there was a booming market in self-help for the technologically confused, including Alvin Toffler's *Future Shock*, with the new term "information overload." Robert Pirsig's bestseller, *Zen and the Art of Motorcycle Maintenance*, offered a new way of thinking about the relationship between human beings and technology, drawing on both Eastern and Western philosophy

Today there's plenty of chatter about the burden of screens, but not the same kind of critical, constructive engagement. Narcissus? Hot and cool media? Does anyone entertain thoughts remotely like this as they scroll through the inbox, vaguely wondering why their mind feels tapped out? We shrug and accept this as our fate. Instead of celebrity philosophers we have celebrity chefs, dozens of them. But they never talk about how delicious *life itself* could be if we followed a different recipe. That's what McLuhan was all about, really, recognizing that the kitchen of the mind is stocked with all the best ingredients. Each of us could be in there every day, cooking up a masterpiece. Why aren't we?

PART III

IN SEARCH OF DEPTH

Ideas in Practice

Chapter Twelve

NOT SO BUSY

Practical Philosophies for Every Day

Thus far into this new era, we've followed a clear-cut approach: we've set out to be as connected as possible, all the time. For most of us, this was not a conscious decision. We did it without really thinking about it, not realizing there was any choice in the matter.

We did have a choice and still do. And because how we live with these devices is a choice, this conundrum is really a philosophical one. It's a matter of the ideas and principles that guide us. If we continue on the current path, over time the costs of this life will erase all the benefits. The answer, therefore, is to adopt a new set of ideas and use them to live in a more thoughtful, intentional way.

There are clues all around us. Whenever I open a gap between myself and my screens, good things happen. I have time and space to think about my life in the digital realm and all the people and information I encounter there. I have a chance to take the outward experiences of the screen back inward. This happened in a small but memorable way the day I called my mother en route to the airport. It was just a routine call, until I put the phone down. Only then did the experience take on unexpected richness and significance.

Such gaps also allow our awareness to return to the physical world. I'm not just a brain, a pair of eyes, and typing fingers. I'm a person with a living body that moves through space and time. In letting screens run my life, I discount the rest of my existence, effectively renouncing my own wholeness. I live a lesser life and give less back to the world. This problem is not just individual and private; it's afflicting all our collective endeavors, in business, schools, and government and at every level of society. We're *living* less and *giving* less, and the world is the worse for it.

This is the moment, while the digital age is still young, to recoup these losses, to bring "all that is human around us," in Google chairman Eric Schmidt's words, back into the equation.

With that aim, in Part II, I went back into the database of human experience in search of helpful ideas. As the seven philosophers showed, this conundrum is as old as civilization. As human connectedness advances, it always makes life busier, by creating new crowds. And life in the crowd inevitably gives rise to the questions we're asking right now: Why don't I have time to think? What's this lost, restless feeling I can't seem to shake? Where does the crowd end, and where do I begin? What are these tools doing to us, and can we fix it?

The philosophers offered all sort of answers, and a number of themes emerged. The most important was the need to strike a healthy balance between connected and disconnected, crowd and self, the outward life and the inward one.

One might argue that civilization always survives such transitions and moves on, so why worry? Of course we'll survive. The question is whether we'll do more than that. In all the earlier periods we've looked at, there were people who thrived and found happiness and people who didn't. The former found something approximating the happy equilibrium Socrates was seeking when he prayed that his outward and inward selves might "be at one." The latter became hostage to their outwardness and never shook "the restless energy of a hunted mind."

Below is a review of the key points along with more concrete ideas about how they might be applied today. The examples are drawn primarily from my own life and experience, because that's what I know. These are suggestions, not prescriptions. Everyone's circumstances are unique, and there's no best approach to this challenge. The purpose of this exercise is to help you develop strategies of your own. Awareness is half the battle, and *any* effort, no matter how small, counts as progress.

1. Plato

Principle: Distance

In Plato's story, Socrates and his friend put the busyness of Athens behind them just by taking a walk. Physical distance is the oldest method of crowd control. In one obvious sense, today it's much harder to go outside the "walls" of the connected life. Truly disconnected places are increasingly rare. But in another way, it's easier. Take a walk without a digital gadget, and distance is yours. The moment you leave all screens behind, you're outside the walls.

Why isn't this a common practice already? Because taking a mobile along seems so harmless and, indeed, sensible. We have acquired a sense that it's dangerous to venture out without one, as though we could never fend for ourselves. It's *nice* to have your digital friend along with you, just in case.

In subtle but important ways, however, it changes the nature of the experience. Though a smart phone brings convenience and a sense of security, it takes away the possibility of true separateness. It's a psychic leash, and the mind can feel it tugging. That's the problem: we've gotten so used to the tug, it's hard to imagine life without it.

To create the modern equivalent of ancient distance and enjoy the benefits it brings, you have to put screens out of reach. Leave the phone in a drawer and walk out the door. Nothing bad is going to happen, and something good just might. Though your disconnected walk might not produce a Socrates-style rapture, it will yield a new sense of inner freedom. Strolling along a city street surrounded by people bent over screens, just knowing *you're* going "commando" puts a spring in your step.

The same underlying principle can be applied to other everyday experiences. Any quick journey out into the world, even the most mundane errand, can double as a miniescape, as long as you have no screen. At the other extreme, try the extended version: an out-of-town holiday. Put on your vacation auto-reply, leave all connective devices at home, and resolve not to check once, even if the opportunity presents itself. Pick a destination, grab a companion, and make a digital escape. If there's a screen at the inn, give it a wide berth.

A few winters ago, *Condé Nast Traveler* magazine sent three reporters to Moscow, one equipped with a BlackBerry, one with an iPhone, and one with just a hard-copy guidebook. They were given a series of tourist challenges to complete in the frigid metropolis, such as finding a great cheap restaurant and locating a pharmacy open at midnight. The low-tech contestant won. After the article ran, one reader wrote in: "I have traveled successfully around the world armed with nothing more than a dog-eared guidebook and a friendly smile. . . . As any seasoned traveler will tell you, the kindness of strangers can be relied upon anywhere. Just don't be too absorbed in your BlackBerry to notice."

Meanwhile, distance in the old-fashioned sense hasn't completely lost its meaning. There *are* still places where it's hard or impossible to find a digital connection of any kind, including remote parts of the continental United States. Take every opportunity to enjoy them, because they won't be around forever. In my family, when we're considering vacation possibilities and summer camps, we perk up when we hear there's no mobile phone or Internet service. Though it's increasingly common for airplane flights to have wireless Internet, not all do. If there's a fee for the service, save your money. You'll be getting a much more valuable amenity—distance from your own connectedness—for free.

2. Seneca
Principle: Inner Space

When physical distance either wasn't available or didn't do the trick, Seneca found inner distance. He did so by focusing on one idea or person and tuning out the rest of the world. Today minimizing the crowd is an even more essential skill, and there are more ways to practice it. The first and most obvious is to choose a friend or family member in your physical vicinity and just have a conversation. A focused, undistracted chat, without screens. It's so obvious, it seems absurd to recommend it. But are we *really* talking to each other anymore? If the person you've focused on has a screen, gently ask him or her to put it aside. What you'll be saying, in effect, is: *I want to be with just you*. It's a rarely heard sentiment these days, and it shouldn't be.

Though letter writing is a dying art, there are plenty of other activities that afford the easygoing absorption of the "flow" state. Especially helpful is anything that involves working with the hands, such as splitting wood, knitting, cooking, or tinkering with a car engine or a bicycle.

We can also minimize the crowd right on the screen, and though it won't bring the inner distance that happens offline, it can help. How many Web pages and other windows do you keep open on your screen at a time? Do you shop online while instant messaging while composing e-mails while randomly checking out videos while playing a game on the side? Try the opposite approach: limit yourself to one screen activity at a time, and don't use the screen to wander away from a phone chat. The person on the other end is to you as Lucilius was to Seneca.

Another strategy for reducing time online is to start using *other people* as your search engines. Rather than constantly checking for news and updates, I let friends and family tell me what's happening. What are the headlines? Which movie star is in trouble? What's the latest outrage on the political front? It's more enjoyable listening to the latest developments through the interpretive lens of a person you know, and it saves a lot of trouble.

Somehow, we've gotten it into our heads that the best use of social-networking technologies is to acquire as many friends and contacts as possible, jamming everyone we know into the same virtual space. Thus, that barely remembered "pal" from elementary school who resurfaced a few weeks ago gets to mix and gossip with our current friends from the office— great.

Back when the Internet was a thrilling novelty, there was a natural tendency to make the most of it by constantly expanding your social connections. Now that much of the human race is online, it makes sense to move in the other direction. Whenever possible, narrow and refine the crowd. While I was writing this book and trying not to be needlessly distracted, I had just one active social network, dedicated solely to a small group of people (less than two dozen) whom I knew during one brief but important period of my life—*and no one else*. Of course, there are endless ways to form smaller groups within online networks, and you don't want to overdo it. Too many subgroups becomes as complex as too many individuals. But, if used intelligently, this tactic can reduce the digital horde to more manageable slices. Rather than firing up my screen and being confronted by everyone I ever knew, when I went to my micro-network, a more intimate group was always waiting for me. *Ah, here's the old gang.* It was the screen equivalent of a neighborhood pub.

3. Gutenberg
Principle: Technologies of Inwardness

Gutenberg made one of the great tools of inwardness, books, available to more people. Could today's technological innovators pull off an equivalent trick with the devices of this moment? The need for inwardness is as great, if not greater. Yet now all the momentum in technology is in the opposite direction, toward *more* intense connectedness, increasing our exposure to the crowd. "All your applications. All at once" said the ad for one handheld, as if "all at once" were helpful to the mind.

The e-book experience is moving in the same direction. Though often touted as a giant step forward, some e-readers are designed to make the experience of reading more outward. Effectively minicomputers with built-in e-mail and Web browsers, they make it much harder to go inward as a reader. Do we really want to make our books as busy as the rest of our lives?

The Gutenberg principle could be applied to many other digital devices, including the notebook computer. If I want to shut out distractions and really get some work done on my notebook, I turn off the wireless, transforming the computer into a disconnected tool. Unfortunately, on my notebook this is a somewhat cumbersome process involving multiple keys. Digital technologies should acknowledge in their design that it's sometimes good to be disconnected. A small but helpful fix would be to provide a prominent Disconnect button that would allow the user to go back and forth easily between the two zones, connected and not. Today, as in the fifteenth century, everyone needs time away from the crowd. Technology should serve that need.

4. Shakespeare
Principle: Old Tools Ease Overload

In the early print era, handwriting didn't go out of style, it came on strong. As Hamlet's "handheld" shows, old tools can be an effective way to bring the information overload of new ones under control. Today older technologies continue to ground the busy mind.

Paper is the best example. Since the middle of the twentieth century, futurists have been predicting the imminent demise of paper. It hasn't happened, because paper is still a useful tool. It's arguably becoming *more* useful, since it offers exactly what we need and crave, a little disconnectedness. Read a paper book. Keep a journal or just jot notes in a simple notebook, as I do in my Moleskine. Subscribe to a new magazine. In a multitasking world where pure focus is harder and harder to come by, paper's seclusion from the Web is an emerging strength. There's nothing like holding a sheaf of beautifully designed pages in your hands. The whole world slows down, and your mind with it.

Don't assume that the newest tools are the best choice for a given task. One year at Eastertime, our son decided to make a drawing for the family gathering at my mother's house. Since he wanted to print a copy for everyone, he headed straight for his iMac and a drawing program called Kid Pix. Wait a second, we said. If he did it at the kitchen table by hand with colored markers, he'd have a lot more artistic freedom. Then he could copy it on his color printer. (He'd also be away from Internet temptations, but we didn't mention that.) He thought about it for a moment and agreed that markers are more fun and expressive. It came out beautifully, and he proclaimed, "Kid Pix isn't very good, anyway."

Old tools are plain fun. As virtual life weighs down on us, material objects paradoxically begin to seem light and playful. Vinyl records not only *do* sound better, they're fascinating to handle and ponder. I take yo-yo breaks in my office. Dominoes and marbles have become a draw. Board games can be bliss.

5. Franklin

Principle: Positive Rituals

Ben Franklin brought order to his chaotic life with a ritual based on positive goals. While he was shooting for "moral perfection," we can aim for the more modest goals of clarity and calm. I've already discussed workplace applications of the Franklin approach, but it applies equally to private life, where there are endless possibilities for finding balance through rituals. Rather than just restricting your own screen time, set time limits and rewards. Somehow, when the battery is running down on a laptop, it's much easier not to be distracted from the task at hand. This behavioral fact can be translated into a ritual. Vow to finish all screen tasks by a given time, with a reward if you make it. You'll get more done, reduce your connected time, and earn a bonus.

Another approach is to keep certain hours of the day screen-free. In *The Tyranny of E-mail*, John Freeman recommends not checking your e-mail early in the morning or late at night, practices he rightly notes create a "workaholic cycle." Of the morning in particular, he writes, "Not checking your e-mail first thing will also reinforce a boundary between your work and your private life, which is essential if you want to be fully present in either place."

Indeed, rituals aimed at offsetting one's digital life don't need to be explicitly *about* digital devices at all. They can be entirely about the positive alternatives. If you've noticed that too many of your evening hours are given over to the screen, resolve to do something completely different and appealing with half of those hours—spend more time with your spouse or partner, study the constellations with a child, or take that Italian cooking course you've been fantasizing about. Design the ritual around the amount of time dedicated to the new positive pursuit, rather than how much you're taking away from the old negative one. Granted, these are just mind tricks, but it's the mind's own unhelpful tricks that we're trying to combat.

6. Thoreau

Principle: Walden Zones

In the middle of the bustling nineteenth century, and relatively close to the crowd, Thoreau created a zone of inner simplicity and peace. Any digital home can serve the same purpose, if properly organized, and there are countless zoning variations. Such spaces don't have to be all about silence and contemplation, which can suggest (especially to children) that offline time is boring. Children should learn that the screen isn't the only place where the action is. If you have a quiet Walden Zone, try to offset it with a loud one, i.e., a space that's both offline *and* rowdy. These can also be established outside the house itself. After all, Thoreau's project was a backyard experiment. Any backyard can be designed as a haven from digital gadgets, a place where the main event is nature itself. The ultimate Walden Zone is a tree house.

As technologies converge toward a future in which one screen will offer all varieties of content—from movies to television to social networking to texting—it might be wise to zone different parts of the home for different *kinds* of screen experiences. Many of us already do this in a de facto way. One room for movies and television-style entertainment best enjoyed in groups, at a distance from the screen, and separate spaces for the close-to-the-screen digital experiences we now associate with computers. It's worth recognizing that these are very distinct activities, which naturally offset each other—the relaxation of television versus the nervousness of keyboard tasks—and it can prove useful to maintain the distinction, so there are clear options within each home.

The Thoreau principle has applications far beyond the private home. There are already Walden Zones in public places—the quiet car on the train is one, but it's about sound rather than screens. Theaters, museums, and some restaurants ask patrons to turn off their devices. Though most schools have been increasing the intensity of their students' connectedness in the last decades, some forward-thinking educators have been creating disconnected environments within their schools for nondigital play and contemplation. Educator Lowell Monke writes that such spaces "give children the opportunity to withdraw from the ceaseless noise of high-tech life and do the kinds of things that their childish nature calls them to do." As long as screens continue to proliferate, this countertrend should only build.

Offline coffee shops? No-screen health clubs? Perhaps a revival of the old Prohibition-era "speakeasy" concept in the form of secret, password-only hangouts for digital fugitives.

7. McLuhan

Principle: Lower the Inner Thermostat

McLuhan said that, even in a busy electronic world, each of us can regulate the quality of our experience. Study the maelstrom that is your busy life, and come up with your own creative ways of escape. An acquaintance of mine cooled down his connectedness by getting rid of his smart phone and returning to a basic cell phone, thus removing e-mails and Internet from his mobile existence. It was "an incredible relief," he says, but there was one problem: he's a huge baseball fan, and losing the smart phone meant he couldn't follow his favorite team on the other side of the country as faithfully. Solution: he found a way to listen to the distant radio coverage via his low-end phone. It not only works beautifully, he reports, but also takes him back to the way he listened to games as a boy.

Our efforts to escape the chaos of digital life don't have to be desperate and arduous. Like Poe's sailor, you can make it a kind of game. "Accidentally" leave your mobile at home when you go out on the weekend, just to see how everyone reacts when they can't reach you. Have a disconnected party where all devices are confiscated at the door. At a chain supermarket where we often shop, digital screens have been installed everywhere, blaring nonstop ads. Sometimes, when nobody's looking, I reach up and flick one off.

Though McLuhan focused on technology over content, the fact is that choosing your content wisely can be a huge help. For instance, having your mind extended out into the world all day in McLuhanesque fashion takes its toll. Thinking globally is exhausting. One way of reining in the overextended mind is to pay closer attention to *local* media content. Instead of always tracking distant happenings, develop a habit of bringing your awareness back home on a regular basis. Make the screen experience less expansive by choosing one good local news site or blog and following it. Listen to local radio channels. Buy a regional newspaper and take it home. Go out and shoot the breeze with a neighbor. The burgeoning "locavore" movement, which promotes consumption of locally raised food, should have a screen equivalent. Escape the global village for your own village, even if it happens to be one square block of a huge city.

And once you *have* that village, here's an idea: organize get-togethers for trading tips about the tools of modern life. At a "SkillShare" event held in our area, people came together to make the digital era a little more collaborative and humane. A story in the next day's newspaper summed it up: "An eighth-grader taught the Nintendo Wii system, two high school boys lectured on Facebook and cellphone features, while a middle-aged man demonstrated how to cut meat." If *that's* a glimpse of the future, we'll all be fine.

The above suggestions are mostly small and incremental, but there are more ambitious ways of applying these ideas. A few years ago, my family and I embarked an experiment aimed at loosening the hold that screens had on our life together. It incorporated some of the ideas discussed above, and it worked so well, it became a permanent feature of our lives. Here's what happened.

Chapter Thirteen

DISCONNECTOPIA

The Internet Sabbath

My family's home life isn't all that different from the lives of our friends in big cities. We're tucked away on a quiet street in a small, remote town, but it's not the study in bucolic disconnectedness it seems. The world has changed dramatically in the last two decades.

Remember the telegraph-era businessman whose mind was "continually on the jump," yanked away from dinner with his loved ones by sudden interruptions from far away? That's what home life is like now for everyone, parents and kids alike. In our case, as our son, William, grew older and developed his own screen interests, more and more it seemed that what we did "together" was the Vanishing Family Trick—we went off to our respective screens.

There's a school of thought that says this is just fine, because digital screens are actually bringing families together. "Technology is enabling new forms of family connectedness that revolve around remote cell phone interactions and communal internet experiences," concluded a study by the Pew Internet & American Life Project, part of the nonprofit Pew Research Center. The study maintained that having multiple computers in one home "does not necessarily lead family members to be in their own isolated technological corners." Rather, it found "many instances where two or more family members go online together, or one calls another over to 'look at this!'"

In other words, the Vanishing Family Trick is even more amazing than it appears. Like the illusionist's lady who disappears from a black box only to materialize on a silken rope lowered from above, the family that disperses at the call of the screen winds up reuniting in a completely different location—on the screen itself! The more we vanish from the fireplace, the closer we become.

But it's not true. My family has been enjoying digital experiences together for years, and they are usually fun and often memorable. I have no doubt that, years from now, one of the moments I'll remember most fondly from William's early years will be three of us sitting at the screen together singing along with the Numa Numa guy, "Numa Numa, hey! Numa Numa, hey!" We gather all the time for music videos, comedy sketches, nature clips, presidential speeches, you name it.

The point isn't that the screen is bad. The screen is, in fact, very good. The point is the lack of proportion, the abandonment of all else, and the strange absent-present state of mind this compulsion produces. "Earth to loved one, are you there? No? Me, neither." We were living for the screen and through the screen, rather than for and through each other

Like the self, the family is a small unit within a much larger crowd, a unit with its own inward life. To flourish and grow, that life requires time apart. Otherwise, both self and family become crowd-dependent, defining themselves in relation to what's *out there* rather than what's *right here*. Thoreau said that the man who goes desperately back to the post office over and over hasn't heard from himself in a long while. The more the members of a family go back to the screen, for whatever it is they seek there, the less they truly hear from one another and the weaker their life together grows.

"Look at this!" is fine. We all enjoy shared spectatorship. But a family isn't a spectator sport. It's all about participation, engagement, connection of the most intimate kind. At our screens, we're all facing outward.

The question was how to turn ourselves around. One option was to configure the physical spaces in the house to create physical Walden Zones dedicated to disconnectedness. In our case, we already effectively had such a zone—it was rare that anyone used a digital device in the living room—but we were being pulled out of it nonetheless. We needed something more comprehensive.

In setting up our home lives, most of us focus only on the physical spaces. We don't give much thought to the temporal dimension, how time will be organized. But we inhabit time together, too, and time can be shaped to serve our needs and goals.

We'd been doing some home renovating, and for inspiration I'd read *A Pattern Language*, a classic architecture and design book from the 1970s written by an architect-philosopher named Christopher Alexander and a few coauthors. The book's basic premise is that there are patterns in the way people around the world have built houses and communities through history. These patterns recur over and over across different cultures and epochs because they reflect deep human desires and needs.

One pattern is Alcoves. A room with alcoves allows a family or any other group to be physically together, while offering each individual the opportunity to be partly by him- or herself, in an alcove. Another recurring pattern, called "Private Terrace on the Street," applies the same need for balance to the exterior of a dwelling:

We have within our natures tendencies toward both communality and individuality. A good house supports *both* kinds of experience: the intimacy of a private haven *and* our participation with a public world. But most homes fail to support these complementary needs. Most often they emphasize one, to the exclusion of the other: we have, for instance, the fishbowl scheme, where living areas face the street with picture windows and the "retreat," where living areas turn away from the street into private gardens.

We were living in a fishbowl arrangement ourselves, I realized. Rather than picture windows, it was our screens that faced the world, tilting the balance toward the crowd. To solve this problem architecturally, the pattern people have settled on through the ages is a terrace configured so one can watch the street from a position of relative privacy. Homes in many traditional cultures have some version of a private terrace facing the street, and they also figure in modern homes. When Frank Lloyd Wright designed a house for a busy street, he sometimes gave it a front terrace with a masonry wall high enough to provide a sense of separation.

As an antidote to our fishbowl, couldn't we use *time* to turn the whole house into a private terrace, where we could be together in a more private, inward-focused way, while still not being completely cut off from the world? Martha and I decided to try a simple experiment, based on the traditional notion of the weekend as time apart. We would turn off the modem at bedtime Friday night and leave it off until Monday morning. Thus on Saturday and Sunday all three family computer screens would be disconnected.

Though this felt like a radical step, it wasn't as if we were going truly off the grid. We would still have our mobile phones.

Neither of us used our phone much for e-mail or Internet, for which they had limited capability anyway (they weren't true smart phones), and we agreed to keep it that way. We did use them to text, but we'd never been texting maniacs. The television would remain plugged in, which we knew wouldn't be a problem. For us television had always been a mostly communal experience, a way of coming together rather than pulling apart.

The modem was the true conduit of our crowd life, the digital water main, and for two days a week it would now be off. We agreed to stick to this plan for several months and see what happened. It would be a positive, Franklinesque ritual, in that we were focused not on what we were giving up but on the benefit we hoped to gain, a more cohesive family life.

We called it the Internet Sabbath. "Ye shall kindle no fire throughout your habitations upon the sabbath day," says the Book of Exodus, and that's basically what we were doing with our screens. They might still be glowing, but without a connection they wouldn't be much of a draw.

The beginning was hard. That first Saturday morning, we woke up in a place that looked just like home but seemed altered in some hard-to-express way. It was as if we'd landed on another planet where the aliens had built a perfect replica of our life, but it was just a stage set and we knew it. Something wasn't right. When you've been in screenworld for a long time, you really lose touch with the third dimension. The rooms were so still and silent, everything in them frustratingly inert and noninteractive. I could feel my mind crawling the surfaces of things, looking for movement, novelty, feedback. Why isn't that coffee table searchable? We were all jonesing for the digital juice and repeatedly found ourselves heading off to our respective corners, only to remember that there was no point.

Beyond the mental adjustments, there were logistical issues.

We told friends and professional contacts that from now on, if they e-mailed us on the weekends, they wouldn't be reaching us right away. If it couldn't wait until Monday, they should call. Some were surprised and intrigued by our scheme; a few were incredulous. Since we both worked from home, we were shutting off the flow of information not just into our family life but into our work spaces as well. How would we get by?

The rules specified that if we really needed something from the Internet, we could go into town and use the public library's computers. The point was to keep the house itself disconnected. We did wind up using the library terminals occasionally, especially in the early months. Later, we got into the habit of anticipating digital needs ahead of time and taking care of them on weekdays. If someone we care about had a birthday coming up, for instance, we made a note to set up the e-card in advance. When a school project was due on Monday, the online research had to be done by Friday evening. In short, we learned to be a little more organized, an unexpected fringe benefit.

Still, we all grumbled. There were lots of specific things we missed right away. No spontaneous Googling to look up a needed fact. No online bill paying, a job I used to perform almost entirely on the weekends. No instant retrieval of driving directions or movie times. If William had a team sports event and it was raining, we couldn't check our inboxes to see if it had been canceled. He missed his online gaming sites. Martha missed e-mail more than I did. I couldn't listen to Internet radio and especially lamented the loss of a certain jazz station out of LA.

But as the weeks and months passed, these went from genuine annoyances to minor inconveniences to nonissues. We'd peeled our minds away from the screens where they'd been stuck. We were really there with one another and nobody else, and we could all feel it. There was an atmospheric change in our minds, a shift to a slower, less restless, more relaxed way of thinking. We could just *be* in one place, doing one particular thing, and enjoy it.

Now and then, we made exceptions. One weekend several months into the regime, a hurricane was bearing down on the Cape and family members were calling to see if we were evacuating. Should we plug back in to track the storm or sit it out and hope for the best? Not a hard call: we plugged in.

One Saturday night, William and I were watching the classic 1950s horror movie *The Blob* through our cable provider's on-demand service, when disaster struck. With eight minutes left in the movie and Steve McQueen trapped in a diner that the Blob was about to swallow whole, the screen went dark. When we tried to start the movie again, we found it had disappeared from the on-demand menu. We called the local Blockbuster, but it didn't have the movie in stock. Our only option was the Internet, where we figured there might be a watchable online version.

So we did it. We broke the Sabbath for the Blob. We found a grainy bootleg that someone had uploaded to YouTube in segments and watched the ending. The need hadn't been urgent by any definition. Under the terms of the Sabbath, it was a grievous sin. But, I told myself, it was done in service of the kind of family togetherness we'd been hoping to encourage. The fact that we'd agonized over it demonstrated to me how far we'd come. Rationalizations, I know. But once we found out what happened to Steve McQueen—phew—we *did* turn the modem back off.

Donald Winnicott, one of the great Freudian psychoanalysts of the last century, wrote an essay called "The Capacity to Be Alone" about how young children develop emotional self-reliance. He said a baby learns to be alone not through true isolation but by being "alone" in the presence of its mother. This occurs when the mother is nearby but not paying close attention to the child. Sensing this, the child begins to grasp its own separateness from her and to understand that it can be alone and still feel protected and safe.

It seems paradoxical that one can learn to be alone by being *with* someone else, but Winnicott contends that the power of aloneness is rooted in this very paradox. Without the existence and knowledge of other people, aloneness would have no meaning. Thus, it's only when the child experiences being alone in this way, with its mother somewhere nearby, that it can grasp the meaning of solitude and embrace its selfhood. Children who don't make this discovery never achieve full maturity, Winnicott said, and lead "a false life built on reactions to external stimuli."

There's a parallel to what the Internet Sabbath did for us. We weren't infants, but we *had* become dependent on external stimuli. And over time that dependency had turned our family life into something that wasn't really us, not the best part of us, anyway. It was a false life, a life less true to our nature as human beings and to the real purpose of family. In turning off the modem, we weren't making the world go away. It was still out there. But like the inattentive mother, it wasn't interacting with us, wasn't making funny hand gestures and goo-goo sounds to keep us stimulated and entertained. In effect, we'd recognized that our screens were infantilizing our life together. And now we were reacquiring, as a family, the "capacity to be alone" that we had lost. It was like growing up again.

None of our online relationships, and nothing else about our digital life, was being sacrificed on the altar of the Sabbath. We were just giving up a series of particular connected experiences that would have unfolded during that forty-eight-hour stretch, almost all of which could take place during the week. The digital medium allows everything to be stored for later use. It was still out there, it was just a little further away. The notion that we could put the crowd, and the crowded part of our life, at a distance like this was empowering in a subtle but significant way. It was a reminder that it was ours to put at a distance. Like Poe's sailor, we had studied the whirlpool and decided that this little move could save us. And it was working.

After six months or so, we reached the point where, instead of dreading the weekly cutoff, we looked forward to it. One Friday night, Martha said she needed a special exception the next morning. There were some urgent work-related e-mails she had to answer before Monday, and she couldn't rely on the library, where the screens are often busy. I skipped my rendezvous with the off switch and went to bed. When I woke up the next morning, I checked to make sure we were still on for a connected morning. "No," she said blearily from under the blankets. "It was so depressing, the thought of waking up on Saturday to e-mail, that I stayed up late and got everything done." I made for the modem, but she'd already taken care of it.

We slowly came to understand, in a visceral way, the high cost of being always connected. At the same time, because we were now away from our connectedness on a regular basis, we grasped its utility and value more fully. We now experienced the two states in an intermittent rhythm, so each could be appreciated in contrast to the other. When I returned to my screen on Monday morning, I was still in a Sabbath state of mind and could do my digital business with more calm and focus, at least for the first couple of days. The inner stillness tended to fade as the week progressed, and by Friday I was ready to go "away" again. A few times, we've sneaked in a spontaneous one-day Sabbath during the week, when one of us needed to clear out the digital fog for an important task.

We've been at it for a few years now, and it's become almost automatic. Sometimes we forget to turn off the modem on Friday and it doesn't make a difference. Having fallen out of the habit of using our screens on those days, it doesn't occur to us to try. An artificially imposed regime has simply become the way we live. On the weekends, the house is a kind of island away from the madness, our disconnectopia. And the good energy we gain from our time there flows over into the rest of life.

This is not to say we're lying around doing nothing. Martha and I still work a lot on weekends, and we keep a full schedule of family activities. The Internet just doesn't figure in any of them. Though digital devices are meant to impose order on our lives, when you remove them, a more natural kind of order returns. It's far easier to be in a room with others and stay there. It's easier to maintain eye contact and have meaningful conversations. It's even easier to be apart from one another. When one of us does drift away from the group, it's to be truly alone with a book or music or just our own thoughts, which now feels healthier. To put it another way, both togetherness and solitude used to be problems for us. Now neither is.

We aren't the only ones who have discovered this. Friends occasionally send us articles and links about others who have tried similar regimens, sometimes calling them Sabbaths. Mark Bittman, a food columnist for the *New York Times*, wrote about a "secular Sabbath" he'd instituted after he checked his e-mail on an airplane flight and realized he was a techno-addict. He'd sworn off for one day a week and now, after six months, was amazed at the transformation: "This achievement is unlike any other in my life." Author Stephen King said it was when he realized he was spending "almost half of each day's consciousness" facing screens that he decided to cut back. "I don't think any man or woman on his or her death-bed ever wished he or she had spent more time sending IMs."

Not every household is in a position to try this. There are jobs and family circumstances that just won't allow for two un-plugged days a week, or even one. Still, I think there are many who could try it without a great deal of inconvenience. When you do something out of conviction, the world has a way of rallying round and lending a hand. I now often get e-mails with subject lines like "I know you won't see this until Monday but . . ." If more people started building alcoves and terraces within the digital environment, new customs and protocols would inevitably arise around them.

Were this idea to spread, it would change the life lived not just inside each home but *between* homes. Another benefit of our Sabbath is that we wind up spending a lot more time outside, seeing the neighbors and enjoying the natural world. According to *A Pattern Language*, the healthiest, most vibrant communities are those in which people meet and mix casually in public squares and other common physical spaces. There's a pattern called "Dancing in the Street," which the book de-scribes as a lost art: "All over the earth, people once danced in the streets. . . . But in those parts of the world that have become 'modern' and technically sophisticated, this experi-ence has died."

Digital society affords a kind of street dancing, through social networks and the like. But it feels more like the Saint Vitus' dance that Thoreau talks about, frantic rather than joyful. If more modems were to shut down on Friday nights, I can see windows being thrown open and people wandering outside the way they do when there's a power outage, meeting neighbors they barely know. There might even be dancing in the streets.

Afterword

Back to the Room

No matter how carefully you think about it or how assiduously you work on new approaches and habits, there's no getting around the fact that we live in a very busy world. So busy that some days you inevitably wind up back in that hectic place where existence is all about tap-tap-tapping, and the very idea of escape feels quixotic.

I had one of those days not long ago. It began with something ostensibly unrelated to digital technology, a summons for jury duty delivered the old-fashioned way, by our mail carrier. When I opened the envelope, my heart sank. Though it's never a good time for a jury call, this one was especially unfortunate. I had a deadline approaching, Martha was working hard on her own book, and various family matters had us both stretched to the limit. And it wasn't the typical order to report to the local courthouse ten minutes from our house, or to the state superior court several towns away. It was a federal jury summons stating that on the appointed morning I was expected to appear at the U.S. District Courthouse in downtown Boston. That meant a two-hour drive each way, possibly more, depending on traffic. I'd be a prospective juror for three weeks, but if I wound up on a jury, it could go much longer.

In the fine print was another unhappy twist: no computers or mobile phones allowed in the building. Based on the ideas I've laid out in these pages, I should have exulted at this out-of-the-blue mandate to spend most of a day disconnected. But to be honest, I didn't. I had a mountain of work to do and much of it required a connected screen. As it is, I get only five digital days a week, and this was a time when I really didn't want to lose one of them. While notching down our connectedness has been fantastic for our family life, and I would never go back, it occasionally presents brand-new dilemmas. In this case, my frustration about not being able to connect during jury duty was mixed with self-loathing about feeling that way. As I lay in bed the night before, pondering all this, I realized I was trapped in the digital room again, and the walls seemed to be closing in.

When I left the house at 4:30 A.M., the waxing moon was enormous in the western sky. Driving along with music playing, I gradually relaxed into a better mood. I had no choice in the matter, after all; might as well make the best of it. Traffic was no problem and I arrived quite early. I parked at the courthouse on Boston Harbor and wandered on foot into the city in search of breakfast. There were few signs of life until, crossing Post Office Square, I noticed overcoated figures converging from all directions on a certain doorway on Milk Street. This being Boston, it naturally turned out to be a Dunkin' Donuts. There was a newspaper vendor out front and, noting that the regulars were all buying his wares, I did the same.

I got a doughnut and coffee, grabbed a stool at the front window next to a young woman submerged in a *Boston Herald*, and opened my *USA Today*. These days, the act of reading a hard-copy newspaper feels curiously out of phase, in two quite different ways. On one hand, it's absurd to be holding these inky sheets in your hands, deciphering words formed by atoms rather than bits. Part of you wonders, *Why on earth am I doing this?* News is supposed to be *new*, and a hard copy is dated long before it arrives. With a screen, you can race around the whole world in seconds, getting all the latest developments in close to real time. The buzz of online news is one of the great pleasures of this era.

On the other hand, a printed newspaper is even more useful now than it was twenty years ago. Like a Moleskine pad, it's a disconnected medium that takes you out of the digital swirl into a calmer, more patient mental space. Buzz is good and important, but so is de-buzzing. There we were, just the pages and me. I could browse lazily, pause over anything that caught my eye, and take the time to think about it, as I rarely do on screen. In this high-speed world, a physical newspaper is a still point for the consciousness. It's also a reminder that *any* room, even a humble doughnut shop, can be a kind of refuge if you know how to use it. It didn't hurt that at this early hour there was no action whatsoever coming from the mobile in my pocket.

"We're Killing Communication," said one headline on the op-ed page. The column was a hilarious rant against all things digital by Bill Persky, a seventy-eight-year-old television writer, producer, and director who had been spending a lot of time with the latest technologies, including social networks. This had brought him a barrage of new "friends" he didn't need and updates about their lives he didn't want, such as "Eating leftover lasagna" and "Getting a colonoscopy." Now, Persky announced, he was quitting the whole scene.

I'm not losing my patience but my sanity. With the wisdom I have gained from age and experience, I have finally decided it's time for all these breakthroughs to take a break from breaking through, since they're no longer improving communication but actually destroying it. How? By making it easier and faster for people everywhere to be in constant contact with each other— about nothing.

I knew just what he was talking about, and I also knew that he was overreacting. Like the mob attacking the printing press in Shakespeare, in his frustration he perceived only the oppressive downside of the new tool, not its many benefits. It's a natural reaction when you feel cornered and see no way out. But, as I could see clearly now from my window seat on Milk Street, there *are* ways out, and they're all around us.

When I got to the courthouse, the other prospective jurors were arriving. Inside the front door was a checkpoint where we had to surrender our phones and computers to armed security officers. There were seventy-five of us in the jury pool that day, but even as the waiting area—an open space with spectacular harbor views—filled up, it remained quiet and still. A crowded room is different when the rest of the world is out of reach. The usual ringtone-initiated yak sessions weren't breaking out every few minutes. Some of us struck up casual conversations with one another, while others read books and paperwork or just stared out at the boats and seagulls. We were *present* in a way people are seldom present anymore.

Many of us had come here reluctantly, convinced we had far more urgent things to do. Under normal circumstances, we would have been spending this time toiling in offices, schools, hospitals, restaurants, and other settings, at tasks we felt deserved our full attention at least as much as jury duty. But if we were in those places now, would we really be giving those tasks our full attention? Doubtful. Too often, devices like the ones we'd left downstairs would be calling the shots, interrupting, distracting, and generally ensuring that our minds never *quite* settled down.

I wasn't placed on a jury, and by midday I was free to go. When I retrieved my phone at the front door, there was a handful of new messages waiting, and I went through them before driving off. Nothing urgent had happened in my absence—and how often does it, really? I'd expected to be in a foul mood now, frantic to make up for lost time. But sitting in those disconnected courthouse rooms had been as refreshing as a long walk in the woods. I'd done some useful thinking and stumbled on several promising new ideas, and I was eager to get back to work.

Experiences like this are crucial, and finding them shouldn't require federal regulations and armed guards. As life in the digital room grows ever more intense, I see a dawning awareness of this need. Not long after jury day, my academic friend—the one who had complained about her nervous system being "interlinked" to those of her colleagues—e-mailed me a news story about colleges "encouraging technology-free introspection." Stephens College in Missouri has revived a long-dormant tradition of vespers, an evening chapel service, but with a digital-age twist. Where the old vespers gatherings were religious, the new ones are secular, expressly designed as contemplative time away from digital devices. Smart phones and other devices are placed in collection baskets so the students can just sit quietly in the pews for an hour. The president of the women's college "fears all that time spent in the twenty-first century's town square leaves few opportunities for clutter-free thought," and she wants these young women to learn self-reliance. Amherst College in Massachusetts organized a "Day of Mindfulness" to give students, in the words of one professor, "a complement to the very hurried world of gadgets they normally live in."

A week later, one of our local newspapers reported on an effort to restore dilapidated houses in our area that are considered historically significant for their twentieth-century modern designs. Many are in isolated spots in the woods, some overlooking tranquil ponds. A nonprofit organization has raised money and begun turning them into places where artists and scholars can live and work for a few weeks at a time. The first artist to stay in one of the houses, a woman from nearby Provincetown, said of the experience: "It's nice to be away from the Internet."

Technology makes the world feel smaller than it really is. There are all kinds of rooms in all kinds of places. Every space is what you make it. But in the end, building a good life isn't about where you are. It's about how you decide to think and live. Place your index finger on your temple and tap twice. It's all in there.

Acknowledgments

Having written so many words about autonomy and self-sufficiency, I have a confession to make: I could not have done this without the help of some wonderful organizations and people. I first began thinking about these ideas in the fall of 2006, when I was lucky enough to spend a semester as a fellow at Harvard University's Joan Shorenstein Center on the Press, Politics and Public Policy. This book was born of that happy experience. My thanks to Alex Jones, Tom Patterson, Nancy Palmer, Edie Holway, and everyone at Shorenstein for the chance to explore and the unwavering support.

I first learned about Hamlet's tables at a marvelous exhibit at the Folger Shakespeare Library in Washington, D.C., where Gail Kern Paster and Heather Wolfe have since helped me in more ways than I can count.

The Huntington Library in San Marino, California, allowed me to spend many productive hours among its rich resources and quiet spaces.

In the fall of 2008, I spent three weeks at the MacDowell Colony in Peterborough, New Hampshire, thinking, writing, and enjoying the company of terrific people. The friendships and encouragement I found there sustained me long afterward.

At a time when striking a healthy balance between solitude and community is such a challenge, MacDowell could be a role model for the world.

When the book was still a work in progress, I was honored to speak at the Woodberry Poetry Room at Harvard, a special place run by a special person, Christina Davis.

This project really began with Dan Okrent, whose support, insights, and friendship have carried me through. Many thanks to Christopher Chabris, Katy Chevigny, Rob Corrigan, Jeffrey Cramer, Tony Horwitz, Walter Isaacson, and Elsa Walsh for reading all or part of the manuscript and offering thoughtful suggestions and comments.

For advice, instruction, conversation, leads, and other kindnesses, I'm grateful to Nicholas Basbanes, Claudia Bedrick, Clara Bingham, Emily Bingham, Dan Bloom, Janis Brennan, Geraldine Brooks, Flip Brophy, David Del Tredici, Bryan Dickson, Tom Djajadiningrat, Nora Gallagher, Howard Gardner, Terry Hanrahan, Asa Hopkins, Sharon Howell, Maxine Isaacs, John Jackson, Constance Kremer, Don Krohn, Becky Okrent, Kees Overbeeke, Julie Piepenkotter, Moin Rahman, Stephen Reily, Samara Sit, Sally Bedell Smith, the Spiegel and Luddy families, Barbara Feinman Todd, Helen Miranda Wilson, John Wolcott, Maryanne Wolf, Tim Woodman, Bob Woodward, and Theo Zimmerman.

Thank you to Jonathan Burnham at HarperCollins, for believing in this idea from the start.

My editor, Gail Winston, is a writer's dream: wise, sensitive, and steadfast. She was a joy to work with and learn from. Any strengths that might be found in these pages are there because Gail pulled them out of me.

To Jason Sack, thank you for great deadline patience and many other assists.

My agent, Melanie Jackson, watched over me as I felt my way through the unfamiliar country of book writing. I could not have asked for a better guardian.

Thank you to my parents and everyone in my family, near and far, and all the friends and neighbors who cheered me on. Your faith in me means more than I can say.

Finally, to Martha and William, thank you for your unflagging support, your inspirations and edits, your sense of humor about all those late nights, and, above all, your love. You are my everything.

Notes

CHAPTER 1: BUSY, VERY BUSY

10   *the "movie-in-the-brain"*: Antonio R. Damasio, "How the Brain Creates the Mind," in *Best of the Brain from Scientific American*, ed. Floyd E. Bloom (New York: Dana Press, 2007), pp. 58–67.

12   *"It all depends"*: William James, "On a Certain Blindness in Human Beings," *On Some of Life's Ideals* (New York: Henry Holt, 1912), p. 37.

13   *William James acknowledged*: Ibid., pp. 3–46.

15   *A news story*: "Teen Tops More than 300,000 Texts in Month: Sacramento Teen Says She's Popular," www.ksbw.com, posted May 5, 2009.

17   *New findings are released*: "Americans Spend Eight Hours a Day on Screens," AFP (Agence France-Presse) wire story, March 27, 2009. Tim Gray, "Study: U.S. Loaded with Internet Addicts," www.sci-tech-today.com, October 18, 2006. "Texting and Driving Worse than Drinking and Driving," www.CNBC.com, June 25, 2009.

CHAPTER 2: HELLO, MOTHER

25   *"Get Connected!"*: *Parade*, November 18, 2007.

25–26 *"ridiculously easy group-forming"*: Clay Shirky, *Here Comes Everybody: The Power of Organizing Without Organizations* (New York: Penguin Press, 2008), p. 155.

28 *an arresting television commercial*: Commercial broadcast on television, fall of 2007.

30 *"I've been to enough Steve Jobs"*: Michael Arrington, "I Am a Member of the Cult of iPhone," www.techcrunch.com, June 10, 2008.

30 *Here was a device*: Frank Bruni, "Where to Eat? Ask Your iPhone," *New York Times*, July 16, 2008; Heidi N. Moore, "Can the iPhone Really Save America?" *Wall Street Journal*, http://online.wsj.com, July 17, 2008.

31 *"tribal experience"*: John Boudreau, "IPhone 3G: 'Worth the Wait,'" www.mercurynews.com, July 12, 2008.

31 *The total number*: Based on data from the International Telecommunication Union (www.itu.int) and *The World Fact Book* (Central Intelligence Agency, www.cia.gov, updated as of November 2009).

31 *surveyed people in seventeen countries*: "The Hyperconnected: Here They Come!," www.idc.com, 2008.

33 *Countries are engaged*: OECD statistics on broadband growth and penetration, www.oecd.org.

33 *Barack Obama*: The campaign quotation about broadband penetration and the postelection address are widely available online.

34 *One reason South Korea*: OECD statistics, www.oecd.org. On gaming as obsession see, for example, the discussion of Seoul as the most connected city on earth, "Most Connected Cities," www.dailywireless.com, March 6, 2007; and "South Korea's Gaming Addicts," BBC News online, November 22, 2002.

CHAPTER 3: GONE OVERBOARD

40 *I'd read a self-help book*: Mildred Newman and Bernard Berkowitz, with Jean Own, *How to Be Your Own Best Friend* (New York: Ballantine Books, 1971); quote appears on pp. 56–57.

42 *Paul Tillich once wrote*: Paul Tillich, *The Eternal Now* (New York: Charles Scribner's Sons, 1963), pp. 17–18.

43 *E. B. White*: E. B. White, *Here Is New York* (New York: Harper & Brothers, 1949), p. 13.

50 *Alvin Toffler coined*: Alvin Toffler, *Future Shock* (New York: Bantam Books, 1984).

50 *According to Edward Hallowell*: Sonja Steptoe, "Q & A: Defining a New Deficit Disorder," www.time.com, January 8, 2006.

51 *Continuous partial attention*: Phrase coined by technology expert Linda Stone; quoted definition is from www .wordspy.com. "E-mail apnea," also attributed to Stone, is defined in "The Too-Much-Information Age," www .yankelovich.com, August 4, 2008.

51 *Internet addiction disorder*: In the March 2008 issue of the *American Journal of Psychiatry*, Dr. Jerald J. Block wrote, "Internet addiction appears to be a common disorder." Block's claim was widely reported. See, e.g., Andy Bloxham, "Internet Addiction Is a 'Clinical Disorder,'" www .telegraph.co.uk, June 20, 2008.

51 *nomophobia*: "Nomophobia Is the Fear of Being out of Mobile Phone Contact—and It's the Plague of Our 24/7 Age," www.thisislondon.co.uk, posted March 31, 2008.

54 Time *magazine*: Claudia Wallis, "The Multitasking Generation," *Time*, March 19, 2006.

55 *The Nielsen Company reported*: See, e.g., Katie Hafner, "Texting May Be Taking a Toll," *New York Times*, May 26, 2009.

55   *true headline*: Dave Carlin, "Teen Girl Falls in Open Manhole While Texting," www.wcbstv.com, July 11, 2009.

55   *nature-deficit disorder*: Richard Louv, *Last Child in the Woods: Saving Our Children from Nature-Deficit Disorder* (Chapel Hill, N.C.: Algonquin Books, 2006).

55   *writer Lowell Monke*: Lowell Monke, "Unplugged Schools," *Orion*, September/October 2007, www.orion magazine.org.

57   *By 2009, people over thirty-five*: See, e.g., Claire Cain Miller, "What's Driving Twitter's Popularity? Not Teens," *New York Times*, August 26, 2009.

58   *why you can sit in a café*: The café scenario I use to illustrate how the mind juggles tasks (or doesn't) was inspired by an analogy drawn by Christopher F. Chabris in "You Have Too Much Mail," *Wall Street Journal*, December 15, 2008.

59   *By some estimates*: Studies by Basex, www.basex.com.

60   *It's estimated*: Ibid.

60   *The brain is the most*: Daniel Tammet, *Embracing the Wide Sky: A Tour Across the Horizons of the Mind* (New York: Free Press, 2009), p. 7.

61   *"the sustained attention of the genius"*: William James, "Attention," in *Talks to Teachers on Psychology; and to Students on Some of Life's Ideals* (Charleston, S.C.: BiblioBazaar, 2007), p. 70.

61   *One of the sharpest observers*: Jen Sorensen, *Slowpoke* strip, *Funny Times*, June 2009.

61   *One study by Basex*: www.basex.com.

62   *"the world's greatest challenge"*: From a statement issued by the Xerox Corporation about the founding of the Information Overload Research Group.

63 *"We will be happy to serve you"*: Hand-lettered sign in the
café at the Providence, Rhode Island, Amtrak station.

63 *"a symbolical mental liberation"*: Risto Etelamaki, quoted
but not named on National Public Radio, August 26,
2008; and by name in Agnieszka Flak, "Ants Bite, Phones
Fly in Finnish Summer Bonanza," www.reuters.com,
August 26, 2008.

63 *"You won't find a television"*: Jeryl Brunner, "10 Unplugged
Vacations," www.forbestraveler.com, June 26, 2008.

CHAPTER 4: SOLUTIONS THAT AREN'T

67 *launched a new ad campaign*: Gates-Seinfeld commercial
broadcast on television, fall of 2008.

68 *Wired.com, for instance*: Brandon Keim, "Digital Overload Is Frying Our Brains," www.wired.com, February 6,
2009.

68 *"Warning"*: L. Gordon Crovitz, "Unloading Information
Overload," http://online.wsj.com, July 7, 2008.

69 The Takeaway: Broadcast of May 22, 2009; listener's
audio comment posted at www.thetakeaway.org.

70 *"There's a competitive advantage"*: IBM researcher John
Tang, quoted in Matt Richtel, "Lost in E-Mail, Tech
Firms Face Self-Made Beast," *New York Times*, June 14,
2008, p. A1.

72 *"up to 950 words a minute"*: Randall Stross, "The Daily
Struggle to Avoid Burial by E-Mail," *New York Times*,
April 20, 2008, Sunday Business section, p. 5.

72 *A popular self-help book*: Timothy Ferriss, *The 4-Hour
Workweek: Escape 9–5, Live Anywhere, and Join the New
Rich* (New York: Crown Publishers, 2007), pp. 114–16.

73 *The human brain is wired*: My discussion of the brain and
attention issues benefited greatly from conversations

with Christopher F. Chabris, Ph.D., of the psychology department at Union College.

74 *"dopamine squirt"*: John Ratey, associate professor of psychiatry, Harvard University, quoted in Matt Richtel, "Driven to Distraction," *New York Times*, July 19, 2009, p. A1.

75 *writes psychologist Steven Pinker*: Steven Pinker, "Will the Mind Figure Out How the Brain Works?" *Time*, April 10, 2000.

75 *Antonio Damasio*: Antonio R. Damasio, "How the Brain Creates the Mind," in *Best of the Brain from Scientific American*, ed. Floyd E. Bloom (New York: Dana Press, 2007), p. 61.

76 *technologies of the self*: Michel Foucault, "Technologies of the Self," in *Technologies of the Self: A Seminar with Michel Foucault*, ed. Luther H. Martin, Huck Gutman, and Patrick H. Hutton (Amherst: University of Massachusetts Press, 1988), p. 16.

76 *"Turn off your computer"*: Eric Schmidt's commencement speech was widely reported on in the news media, for example: Kathy Matheson, "Google CEO Urges Grads: 'Turn off your computer,'" Associated Press, May 18, 2009.

CHAPTER 5: WALKING TO HEAVEN
I relied mostly on the translation of Plato's *Phaedrus* by Alexander Nehamas and Paul Woodruff in *Plato: Complete Works*, ed. John M. Cooper (Indianapolis: Hackett Publishing, 1997). All quotes are from this version, except in two places where I preferred Benjamin Jowett's nineteenth-century translation. The Jowett quotes are from *Symposium and Phaedrus* (New York: Dover, 1993). I made these choices based not on the faithfulness of the translation (I don't know ancient

Greek) but on the meaning in English and how it related to my particular topic. Unless I specifically cite Jowett, all citations of *Phaedrus* refer to the Nehamas-Woodruff version. There is one word, "scroll," that does not appear in either of these translations but does in some others, and I use it for the reasons explained below.

83 *a young man*: According to some sources, Phaedrus would have been closer to middle age at the time the actual conversation took place. Since Socrates calls him a "boy," I assume here that he was in fact young.

83 *"Phaedrus, my friend!"*: Plato, *Phaedrus*, trans. Alexander Nehamas and Paul Woodruff, in *Plato: Complete Works*, ed. John M. Cooper (Indianapolis: Hackett Publishing, 1997), p. 507.

84 *"it's more refreshing"*: Ibid., p. 507.

85 *"you never even set foot"*: Ibid., p. 510.

85 *"Forgive me, my friend"*: Ibid., p. 510.

85 *In fact, this is the only one*: John M. Cooper, introduction to ibid., p. 506.

86 *"They invented talking"*: E. H. Gombrich, *A Little History of the World*, trans. Caroline Mustill (New Haven, Conn.: Yale University Press, 2005), p. 7.

87 *"the greatest inventors of all time"*: Ibid., p. 5.

87 *"He denied that he had discovered"*: John M. Cooper, introduction to *Plato: Complete Works*, p. xix.

90 *Some translations*: Though both the Alexander Nehamas and Paul Woodruff translation and the Benjamin Jowett use "book," other translations use "scroll." I prefer the latter because the word "book" brings to mind the familiar codex of our own time, which wouldn't be invented for several hundred years.

91 *"the place beyond heaven"*: *Phaedrus*, p. 525.

91 *"is inevitably a painfully"*: Ibid., p. 524.

91 *"trampling and striking"*: Ibid., p. 526.

91 *"pure knowledge"*: Ibid., p. 525.

92 *"The result is terribly noisy"*: Ibid., p. 526.

93 *"There's something really divine"*: Ibid., p. 517.

93 *He tells a story*: Ibid., pp. 551–52.

94 *"remember it from the inside"*: Ibid., p. 552.

94 *"[T]hey will be tiresome"*: Plato, *Symposium and Phaedrus*, trans. Benjamin Jowett (New York: Dover, 1993), p. 88.

94 *"stand there as if"*: *Phaedrus*, p. 552.

94 *"continues to signify"*: Ibid., p. 552.

95 *"lovely, pure and clear"*: Ibid., p 509.

96 *"this thing we are carrying around"*: Ibid., p. 528.

100 *"Beloved Pan"*: Plato, *Symposium and Phaedrus*, trans. Jowett, p. 92.

CHAPTER 6: THE SPA OF THE MIND

All quotations from Seneca's letters to Lucilius are from *Letters from a Stoic*, translated by Robin Campbell. After the first reference below, I refer to this book simply as *Letters*.

107 *He was born*: Biographical information about Seneca comes from two sources: Miriam T. Griffin, *Seneca: A Philosopher in Politics* (Oxford: Oxford University Press, 1976), and Robin Campbell's introduction to his translation of Seneca's letters.

107 *"the real master of the world"*: Pierre Grimal, *The Civilization of Rome* (New York: Simon & Schuster, 1963), p. 497.

108 *"It is not the man"*: Seneca, *Letters from a Stoic*, trans. Robin Campbell (London: Penguin Books, 2004), p. 34.

109 *"from all directions"*: *Letters*, p. 125.

110 *"You ask me"*: Ibid., p. 41.

111 *"the restless energy"*: Ibid., p. 36.

111 *"All this hurrying"*: Ibid., p. 189.

111 *"The man who spends his time"*: Ibid., p. 186.

111 *A Roman bookseller*: Harold A. Innis, *Empire and Communications*, rev. Mary Q. Innis (Toronto: University of Toronto Press, 1972), p. 106.

112 *"skip from one to another"*: Letters, p. 33.

112 *"Food that is vomited up"*: Ibid., p. 33.

112 *Seneca tells the story*: Ibid., pp. 73–75.

113 *"Measure your life"*: Ibid., p. 160.

113 *"It's about treating"*: Winifred Gallagher, *Rapt: Attention and the Focused Life* (New York: Penguin Press, 2009), p. 53.

113 *"After running over"*: Letters, p. 34.

114 *"I cannot for the life of me"*: Ibid., p. 109.

114 *"Picture me"*: Ibid.

114 *street sounds:* Ibid., pp. 109–10.

114 *"I swear I no more notice"*: Ibid., p. 110.

115 *"able at will"*: Ibid., p. 186.

115 *"inward detachment"*: Ibid., p. 112.

115 *What Seneca describes*: Mihaly Csikszentmihalyi, *Flow: The Psychology of Optimal Experience* (New York: Harper Perennial, 1991), pp. 2–6 and p. 49.

116 *Queen Elizabeth I*: Robin Campbell, introduction to *Letters*, p. 25, and related note, p. 238.

118 *Gadgets now exist*: One example is a downloadable add-on called Readability, http://lab.arc90.com/experiments/readability.

CHAPTER 7: LITTLE MIRRORS

121 *"Even better"*: Anita Hamilton, "The iPhone: Second Time's a Charm," www.time.com, July 14, 2008.

121 *"the 'Jesus phone'"*: John Boudreau, "IPhone 3G: 'Worth the Wait,'" www.mercurynews.com, July 12, 2008.

122 *"The store's entrance was besieged"*: Connie Guglielmo and Pavel Alpeyev, "Apple's New IPhone Debut Draws Crowds, Helicopters," www.bloomberg.com, July 11, 2008.

124 *Saint Augustine*: Alberto Manguel, *A History of Reading* (New York: Viking, 1996), pp. 41–51.

124 *"oral skill"*: Ibid., p. 47.

126 *In 1432*: My account of the Aachen pilgrimages relies principally on two sources: John Man, *Gutenberg: How One Man Remade the World with Words* (New York: John Wiley & Sons, 2002), and Albert Kapr, *Johann Gutenberg: The Man and His Invention*, trans. Douglas Martin (Aldershot, England: Scolar Press, 1996).

128 *"as if it were"*: Man, *Gutenberg*, p. 63.

128 *"You could head for home"*: Ibid.

128 *"no doubt command"*: Victor Scholderer, *Johann Gutenberg: The Inventor of Printing* (London: Trustees of the British Museum, 1963), p. 10.

131 *A man who saw them*: Manguel, *A History of Reading*, pp. 133–34.

131 *In his recent book*: Robert Darnton, *The Case for Books: Past, Present, and Future* (New York: Public Affairs, 2009), pp. xiv–xv.

133 *"[I]t has proven"*: Ibid., p. 68.

133 *"an early capitalist"*: Man, *Gutenberg*, p. 8.

135 *"Closing the book"*: The quoted lines are from Stafford's poem "An Afternoon in the Stacks." www.williamstafford .org.

CHAPTER 8: HAMLET'S BLACKBERRY
All quotations from Shakespeare in this chapter are from Stephen Greenblatt, general ed., *The Norton Shakespeare* (New

York: W. W. Norton & Company, 1997). Since these are well known, I have not included specific references for each one. All notes citing Greenblatt refer to his Shakespeare biography, *Will in the World*.

140 *"the London crowd"*: Stephen Greenblatt, *Will in the World* (New York: W. W. Norton & Company, 2004), p. 169.

141 *"the livid and decaying heads"*: Mark Twain, *The Prince and the Pauper* (New York: Modern Library, 2003), p. 64.

141 *"the poet of the human race"*: Ralph Waldo Emerson, "Shakespeare; or, the Poet," from *Representative Men* in *Ralph Waldo Emerson: Essays & Lectures*, ed. Joel Porte (New York: Library of America, 1983), p. 717.

142 *In certain cases, accused criminals*: Greenblatt, *Will in the World*, p. 171.

143 *those who lived through*: See Ann Blair, "Reading Strategies for Coping with Information Overload, ca. 1550–1700," *Journal of the History of Ideas* 64 (2003), pp. 11–28; and Blair's forthcoming *Too Much to Know: Managing Scholarly Information Before the Modern Age* (New Haven, Conn.: Yale University Press, 2010).

146 *"collecting pieces of poetry"*: Peter Stallybrass, Roger Chartier, J. Franklin Mowery, and Heather Wolfe, "Hamlet's Tables and the Technologies of Writing in Renaissance England," *Shakespeare Quarterly* 55, no. 4 (2004), pp. 380–419.

147 *Users spoke effusively*: All material about the popularity of tables, including the Montaigne and Sharpham quotations, is from ibid.

148 *"Time has given the hinge"*: Paul Duguid discusses the hinge in "Material Matters: Aspects of the Past and the Futurology of the Book," in *The Future of the Book*, ed. Geoffrey Nunberg (Berkeley: University of California Press, 1996), pp. 63–102.

150 *"The advent of printing"*: Peter Stallybrass, Michael Mendle, and Heather Wolfe, text of brochure for "Technologies in the Age of Print," exhibit at the Folger Shakespeare Library, Washington, D.C., September 28, 2006–February 17, 2007.

151 *"To make cleane your Tables"*: Ibid.

153 *embodied interaction*: On embodiment and technology, I am indebted to the work of Abigail J. Sellen and Richard H. R. Harper, the authors of *The Myth of the Paperless Office* (Cambridge, Mass.: MIT Press, 2003); Moin Rahman of the Motorola Corporation; Tom Djajadiningrat of Philips Design; and Professor Kees Overbeeke of Eindhoven University of Technology, the Netherlands.

155 *"intense representation of inwardness"*: Greenblatt, *Will in the World*, p. 323.

CHAPTER 9: INVENTING YOUR LIFE

For background information about Benjamin Franklin's life and times, I relied on three sources: Franklin, *The Autobiography of Benjamin Franklin*; Walter Isaacson, *Benjamin Franklin: An American Life*; and Carl Van Doren, *Benjamin Franklin*. Any reference specific to one of these books is noted below.

158 *"Withdraw it even for a day"*: Sue Shellenbarger, "A Day Without Email Is Like . . . ," *Wall Street Journal*, October 11, 2007, p. D1.

163 *"One of the fundamental sentiments"*: Walter Isaacson, *Benjamin Franklin: An American Life* (New York: Simon & Schuster, 2004), p. 50.

163 *On the long voyage*: The story of the card cheat and resulting journal entry are from Carl Van Doren, *Benjamin Franklin* (New York: Viking Press, 1938), pp. 61–62.

164 *"I have never fixed"*: Ibid., p. 63.

165 *two fictional dialogues*: Dialogue excerpts are from ibid., pp. 83–87.

167 *"Franklin's powers"*: Ibid., p. 782.

167 *"follow the Example"*: Benjamin Franklin, *The Autobiography of Benjamin Franklin* in *Benjamin Franklin: Autobiography, Poor Richard, and Later Writings*, ed. J.A. Leo Lemay (New York: Library of America, 1997), p. 651.

168 *"He made a list"*: D. H. Lawrence, *Studies in Classic American Literature* (New York: Penguin Books, 1977), p. 17.

170 *Intel has devoted*: Jonathan B. Spira and Cody Burke, "Intel's War on Information Overload: A Case Study," www.basex.com.

172 *"Information overload"*: Jonathan B. Spira, "A Day Without E-mail," www.basexblog.com, December 9, 2009.

173 *"All new tools"*: Franklin, *Autobiography*, p. 378.

CHAPTER 10: THE WALDEN ZONE

180 *"I went to the woods"*: Henry David Thoreau, *Walden* in *Walden and Other Writings of Henry David Thoreau*, ed. Brooks Atkinson (New York: Modern Library, 1937), p. 81.

180 *"I love to be alone"*: Ibid., p. 122.

182 *"The mass of men"*: Ibid., p. 7.

182 *A brand-new track*: Robert D. Richardson, Jr., *Henry Thoreau: A Life of the Mind* (Berkeley: University of California Press, 1986), pp. 137–39.

183 *"learned that his heart"*: Ibid., p. 136.

183 *"It would be some advantage"*: Thoreau, *Walden*, p. 10.

183 *"it was clear to him"*: Richardson, *Henry Thoreau*, p. 153.

184 *"A slender wire"*: "The Telegraph," unsigned editorial, *New York Times*, September 14, 1852; accessed online at www.nytimes.com.

185 *"The merchant goes home"*: Tom Standage, *The Victorian Internet: The Remarkable Story of the Telegraph and the Nineteenth Century's On-line Pioneers* (New York: Berkley Books, 1999), p. 166.

185 *"But lo!"*: Thoreau, *Walden*, p. 33.

186 *"pretty toys"*: Ibid., pp. 46–47.

186 *"As I went under"*: Thoreau, journal entry, September 3, 1851, in *The Heart of Thoreau's Journals*, ed. Odell Shepard (New York: Dover, 1961), p. 57.

186 *"the very best lead pencils"*: Henry Petroski, *The Pencil: A History of Design and Circumstance* (New York: Alfred A. Knopf, 1992), p. 4.

187 *"We are eager to tunnel"*: Thoreau, *Walden*, p. 47.

187 *"Why should we live"*: Ibid., p. 83.

187 *"Surface meets surface"*: Thoreau, "Life Without Principle," in Atkinson, ed., *Walden and Other Writings of Henry David Thoreau*, pp. 723–24.

188 *"the voices"*: Ralph Waldo Emerson, "Self-Reliance," from *Essays: First Series*, in *Ralph Waldo Emerson: Essays & Lectures*, ed. Joel Porte (New York: Library of America, 1983), p. 261.

188 *"My life is superficial"*: Ralph Waldo Emerson, "The Transcendentalist" in *Ralph Waldo Emerson: Essays & Lectures*), p. 205.

189 *"Simplify, simplify"*: Thoreau, *Walden*, p. 82.

189 *"By simplifying our outward lives"*: Thoreau, *Letters to a Spiritual Seeker*, ed. Bradley P. Dean (New York: W. W. Norton & Company, 2006), p. 21.

189 *"I love society"*: Thoreau, *Walden*, p. 127.

189 *"fewer came to see me"*: Ibid., p. 130.

189 *"I went about my business"*: Ibid., pp. 137–38.

189–90 *"I had three chairs"*: Ibid., p. 127.

190 *"So easy it is"*: Ibid., p. 129.

190 *"It is something"*: Ibid., p. 81.

190 *Gandhi*: Richardson, *Henry Thoreau*, pp. 316–17.

190 *"a backyard laboratory"*: Ibid., p. 171.

191 *"All thoroughly satisfactory"*: Dorothy J. Field, *The Human House* (New York: Houghton Mifflin, 1939), p. 17.

192 *"You think that I am"*: Thoreau, journal entry, February 8, 1857, in *The Heart of Thoreau's Journals*, p. 173.

CHAPTER 11: A COOLER SELF

194 *"involuntarily altered"*: Marshall McLuhan, *The Gutenberg Galaxy: The Making of Typographic Man* (Toronto: University of Toronto Press, 1962), p. 183.

196 *"For life today in America"*: Anne Morrow Lindbergh, *Gift from the Sea* (New York: Pantheon Books, 2005), p. 20.

198 *"Man the tool-making animal"*: McLuhan, *The Gutenberg Galaxy*, p. 4.

198 *"inner direction"*: Ibid., p. 28.

199 *"a total field of interacting events"*: Marshall McLuhan, *Understanding Media: The Extensions of Man* (Cambridge, Mass.: MIT Press, 1995), p. 248.

199 *"Technologies create new environments"*: W. Terrence Gordon, "Terrence Gordon on Marshall McLuhan and What He Was Doin'," *The Beaver* 84 (2), May 2004.

199 *"instead of being pushed around"*: McLuhan, *The Gutenberg Galaxy*, p. 6.

200 *"My suggestion"*: Ibid., p. 35.

201 *"Now the point of this myth"*: McLuhan, *Understanding Media*, p. 41.

202 *"is simply in knowing"*: Ibid., p. 15.

202 *McLuhan used*: Edgar Allan Poe, "A Descent into the Maelstrom," in *Edgar Allan Poe: Poetry and Tales*, ed.

Patrick F. Quinn (New York: Library of America, 1984), pp. 432–48.

203 *"How are we to get out"*: Question asked by McLuhan in the film *McLuhan's Wake*, as quoted in Brian D. Johnson, "A Prophet Gets Some Honour," *Maclean's*, December 2, 2002.

203 *"People are cowed"*: Kevin McMahon, quoted in Johnson, "A Prophet Gets Some Honour."

204 *"The hot form excludes"*: McLuhan, *Understanding Media*, p. 23.

204 *a booming market*: Alvin Toffler, *Future Shock* (New York: Bantam Books, 1984); Robert M. Pirsig, *Zen and the Art of Motorcycle Maintenance: An Inquiry into Values* (New York: HarperTorch, 2006).

CHAPTER 12: NOT SO BUSY

212 Condé Nast Traveler *magazine*: "Get Smart? Testing the iPhone and the Blackberry Bold," *Condé Nast Traveler*, June 2009. Follow-up letter from Becca Podell published in the August 2009 issue.

215 *"All your applications"*: Online ad for the Palm Pre.

217 *In* The Tyranny of E-mail: John Freeman, *The Tyranny of E-Mail* (New York: Scribner's, 2009), pp. 208–10.

219 *Educator Lowell Monke*: Lowell Monke, "Unplugged Schools," *Orion*, September/October 2007, www.orion magazine.org.

221 *"An eighth-grader"*: K. C. Myers, "Have a Skill? Please Share!" *Cape Cod Times*, October 4, 2009, p. A1.

CHAPTER 13: DISCONNECTOPIA

223 *"Technology is enabling"*: Pew Internet & American Life

Project, "Networked Families," www.pewinternet.org, October 19, 2008.

225 *One pattern is*: Christopher Alexander, Sara Ishikawa, and Murray Silverstein, *A Pattern Language* (New York: Oxford University Press, 1977), pp. 828–32 and p. 665.

226 *When Frank Lloyd Wright designed*: Ibid., p. 665.

229 *Donald Winnicott*: Donald Winnicott, "The Capacity to Be Alone," in *The Maturational Processes and the Facilitating Environment* (London: Karnac Books, 1990), pp. 29–36.

232 *We aren't the only ones*: Mark Bittman, "I Need a Virtual Break. No, Really," www.nytimes.com, March 2, 2008; "King's Screen Addiction," *The Week*, August 7, 2009, p. 10.

233 *"Dancing in the Street"*: Alexander, Ishikawa, and Silverstein, *A Pattern Language*, pp. 319–21.

AFTERWORD: BACK TO THE ROOM

237 *"We're Killing Communication"*: Bill Persky, "We're Killing Communication," *USA Today*, November 2, 2009, p. 9A.

239 *e-mailed me a news story*: Alan Scher Zagier, "College Asks Students to Power Down, Contemplate," *Washington Post*, December 25, 2009, www.washingtonpost.com.

240 *one of our local newspapers*: Mary Ann Bragg, "Modernist Makeover in Wellfleet," *Cape Cod Times*, January 2, 2010.

Further Reading

This book grew out of an essay that I wrote several years ago as a fellow at Harvard University's Joan Shorenstein Center on the Press, Politics and Public Policy. Also called "Hamlet's BlackBerry," it's about the past, present, and future of one particular tool of human connectedness, paper. The essay is available online at my Web site, www.williampowers.com.

For readers who want to explore further the ideas discussed in this book, below is a list of books that were useful to me in my thinking and research. Not all are mentioned in the foregoing chapters and I don't agree with all of the authors' ideas and conclusions. But for one reason or another, I found each of these books worth reading.

PHILOSOPHY AND EVERYDAY LIFE

Botton, Alain de. *The Consolations of Philosophy*. New York: Vintage, 2001.

Hadot, Pierre. *Philosophy as a Way of Life: Spiritual Exercises from Socrates to Foucault*, ed. Arnold I. Davidson, trans. Michael Chase. Malden, Mass.: Blackwell Publishing, 1995.

James, William. *On Some of Life's Ideals*. Two lectures by James published together in book form during his lifetime. The first, "On a Certain Blindness in Human Beings" explores the challenge of finding "vital significance" in one's daily life. There are various paperback reprints of the book, and the text of this beautiful lecture can also be found online.

Richards, M. C. *Centering: In Pottery, Poetry, and the Person*. Middletown, Conn.: Wesleyan University Press, 1989.

AUTONOMY AND SOLITUDE

Lindbergh, Anne Morrow. *Gift from the Sea*. New York: Pantheon Books, 2005.

Newman, Mildred, and Bernard Berkowitz, with Jean Owen. *How to Be Your Own Best Friend*. New York: Ballantine Books, 1986.

Storr, Anthony. *Solitude: A Return to the Self*. New York: Ballantine Books, 1989.

TOOLS AND PEOPLE

Brown, John Seely, and Paul Duguid. *The Social Life of Information*. Boston: Harvard Business School Press, 2000.

Pirsig, Robert M. *Zen and the Art of Motorcycle Maintenance: An Inquiry Into Values*. New York: HarperTorch, 2006.

Sellen, Abigail J., and Richard H. R. Harper. *The Myth of the Paperless Office*. Cambridge, Mass.: MIT Press, 2003.

PHILOSOPHY OF HOME DESIGN

Alexander, Christopher, Sara Ishikawa, and Murray Silverstein with Max Jacobson, Ingrid Fiksdahl-King, and Shlomo Angel. *A Pattern Language*. New York: Oxford University Press, 1977.

PLATO

Cooper, John M., ed. *Plato: Complete Works*. Indianapolis: Hackett Publishing, 1997.

SENECA

Seneca. *Letters from a Stoic: Epistulae Morales ad Lucilium*, selected and trans. with an introduction by Robin Campbell. London: Penguin Books, 2004.

On Stoicism

Aurelius, Marcus. *The Meditations of Marcus Aurelius*. Widely available in paperback and online.

Epictetus. *Epictetus: Discourses and Selected Writings*, trans. and ed. Robert Dobbin. London: Penguin Books, 2008.

On Concentration

Csikszentmihalyi, Mihaly. *Flow: The Psychology of Optimal Experience*. New York: HarperPerennial, 1991.

Gallagher, Winifred. *Rapt: Attention and the Focused Life*. New York: Penguin Press, 2009.

GUTENBERG

Man, John. *Gutenberg: How One Man Remade the World with Words*. New York: John Wiley & Sons, 2002.

On Books and Reading

Basbanes, Nicholas A. *A Splendor of Letters: The Permanence of Books in an Impermanent World*. New York: HarperCollins, 2003.

Darnton, Robert. *The Case for Books: Past, Present, and Future*. New York: PublicAffairs, 2009.

Manguel, Alberto. *A History of Reading*. New York: Viking, 1996.

SHAKESPEARE

Greenblatt, Stephen. *Will in the World: How Shakespeare Became Shakespeare*. New York: W. W. Norton & Company, 2004.

FRANKLIN

Franklin, Benjamin. *Autobiography.* Widely available.

Isaacson, Walter. *Benjamin Franklin: An American Life.* New York: Simon & Schuster, 2004.

THOREAU

Richardson, Robert D., Jr. *Henry Thoreau: A Life of the Mind.* Berkeley: University of California Press, 1986.

Standage, Tom. *The Victorian Internet: The Remarkable Story of the Telegraph and the Nineteenth Century's On-line Pioneers.* New York: Berkley, 1999.

Thoreau, Henry David. *The Heart of Thoreau's Journals,* ed. Odell Shepard. New York: Dover, 1961.

———. *Letters to a Spiritual Seeker,* ed. Bradley P. Dean. New York: W. W. Norton & Company, 2006.

———. *Walden,* ed. Jeffrey S. Cramer. New Haven, Conn.: Yale Nota Bene, 2006.

MCLUHAN

Gordon, W. Terrence. *Marshall McLuhan: Escape into Understanding.* New York: Basic Books, 1998.

McLuhan, Marshall. *The Gutenberg Galaxy: The Making of Typographic Man.* Toronto: University of Toronto Press, 1962.

———. *Understanding Media: The Extensions of Man.* Cambridge, Mass.: MIT Press, 1995.

Twentieth-Century Thought on Crowds

Canetti, Elias. *Crowds and Power,* trans. Carol Stewart. New York: Farrar, Straus and Giroux, 1984.

Hoffer, Eric. *The True Believer: Thoughts on the Nature of Mass Movements.* New York: Perennial Classics, 2002.

Riesman, David, with Reuel Denney and Nathan Glazer. *The Lonely Crowd.* New Haven, Conn.: Yale Nota Bene, 2001.

Post-McLuhan Thought on Technology and People

Freeman, John. *The Tyranny of E-mail: The Four-Thousand-Year Journey to Your Inbox.* New York: Scribner, 2009.

Lanier, Jaron. *You Are Not a Gadget: A Manifesto.* New York: Alfred A. Knopf, 2010.

Postman, Neil. *Technopoly: The Surrender of Culture to Technology.* New York: Vintage Books, 1993.

Shenk, David. *Data Smog: Surviving the Information Glut.* New York: HarperOne, 1998.

Shirky, Clay. *Here Comes Everybody: The Power of Organizing without Organizations.* New York: Penguin Press, 2008.